CW01621506

EDEXCEL FURTHER MATHS

FURTHER STATISTICS 1

For AS and A Level

Series Editor
David Baker

Authors
Rob Wagner, John Rayneau

Katie Wood, Garry Wiseman, Eddie Mullan, Brian Jefferson, Mike Heylings, David Bowles

OXFORD UNIVERSITY PRESS

OXFORD
UNIVERSITY PRESS

Great Clarendon Street, Oxford, OX2 6DP, United Kingdom

Oxford University Press is a department of the University of Oxford.

It furthers the University's objective of excellence in research, scholarship, and education by publishing worldwide. Oxford is a registered trade mark of Oxford University Press in the UK and in certain other countries.

First published in 2018

British Library Cataloguing in Publication Data
Data available

978 0 19 841527 5

10 9 8 7 6 5 4 3 2 1

Paper used in the production of this book is a natural, recyclable product made from wood grown in sustainable forests.
The manufacturing process conforms to the environmental regulations of the country of origin.

Printed in Great Britain by CPI Anthony Rowe

Acknowledgements
Authors
Rob Wagner, John Rayneau
Katie Wood, Garry Wiseman, Eddie Mullan, Brian Jefferson, Mike Heylings, David Bowles

Editorial team
Dom Holdsworth, Ian Knowles, Matteo Orsini Jones, Felicity Ounsted, Rosie Day, Sarah Dutton

With thanks also to Geoff Wake, Matt Woodford, Katherine Bird for their contribution.

Index compiled by Marian Preston, Preston Indexing.

Although we have made every effort to trace and contact all copyright holders before publication, this has not been possible in all cases. If notified, the publisher will rectify any errors or omissions at the earliest opportunity.

p1: monkeybusinessimages/iStockphoto,
p51: Alphotographic/iStockphoto

Contents

About this book

This book has been specifically created for those studying the Edexcel 2017 Further Mathematics AS and A Level. It's been written by a team of experienced authors and teachers, and it's packed with questions, explanation and extra features to help you get the most out of your course.

Every section starts by covering the basic **Fluency and skills** (A01).

Key points highlight important concepts, and make the information easier to digest.

Worked examples provide a model answer and commentary to realistic practice questions.

There is a Fluency and skills exercise for each section, to practise the skills before moving on to the Reasoning and problem-solving section.

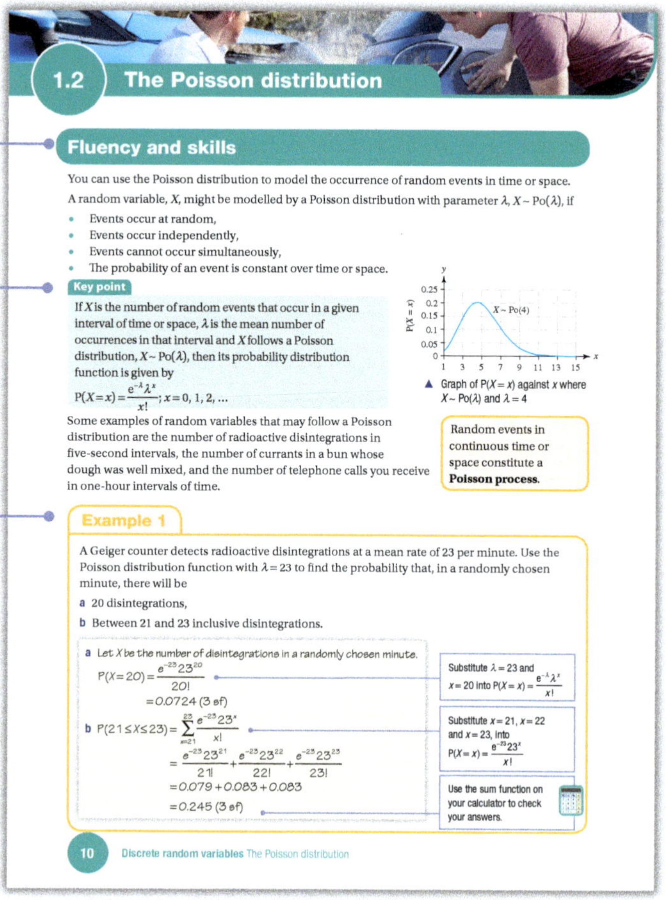

1.2 The Poisson distribution

Fluency and skills

You can use the Poisson distribution to model the occurrence of random events in time or space.

A random variable, X, might be modelled by a Poisson distribution with parameter λ, $X \sim \text{Po}(\lambda)$, if

- Events occur at random,
- Events occur independently,
- Events cannot occur simultaneously,
- The probability of an event is constant over time or space.

Key point

If X is the number of random events that occur in a given interval of time or space, λ is the mean number of occurrences in that interval and X follows a Poisson distribution, $X \sim \text{Po}(\lambda)$, then its probability distribution function is given by

$\text{P}(X=x)=\frac{e^{-\lambda}\lambda^x}{x!};\ x=0,1,2,\ldots$

▲ Graph of $\text{P}(X=x)$ against x where $X \sim \text{Po}(\lambda)$ and $\lambda = 4$

Some examples of random variables that may follow a Poisson distribution are the number of radioactive disintegrations in five-second intervals, the number of currants in a bun whose dough was well mixed, and the number of telephone calls you receive in one-hour intervals of time.

Random events in continuous time or space constitute a **Poisson process.**

Example 1

A Geiger counter detects radioactive disintegrations at a mean rate of 23 per minute. Use the Poisson distribution function with $\lambda = 23$ to find the probability that, in a randomly chosen minute, there will be

a 20 disintegrations,

b Between 21 and 23 inclusive disintegrations.

a Let X be the number of disintegrations in a randomly chosen minute.

$\text{P}(X=20)=\frac{e^{-23}23^{20}}{20!}$ — Substitute $\lambda = 23$ and $x = 20$ into $\text{P}(X=x)=\frac{e^{-\lambda}\lambda^x}{x!}$

$=0.0724$ (3 sf)

b $\text{P}(21\le X\le 23)=\sum_{x=21}^{23}\frac{e^{-23}23^x}{x!}$ — Substitute $x = 21$, $x = 22$ and $x = 23$, into $\text{P}(X=x)=\frac{e^{-23}23^x}{x!}$

$=\frac{e^{-23}23^{21}}{21!}+\frac{e^{-23}23^{22}}{22!}+\frac{e^{-23}23^{23}}{23!}$

$=0.079+0.083+0.083$

$=0.245$ (3 sf) — Use the sum function on your calculator to check your answers.

10 Discrete random variables The Poisson distribution

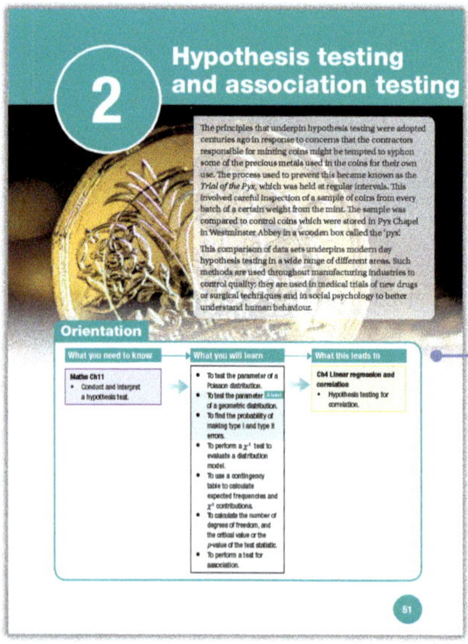

2 Hypothesis testing and association testing

Orientation

What you need to know | What you will learn | What this leads to

On the chapter **Introduction page**, the **Orientation box** explains what you should already know, what you will learn, and what this leads to.

At the end of every chapter, an **Exploration page** gives you an opportunity to explore the subject beyond the specification.

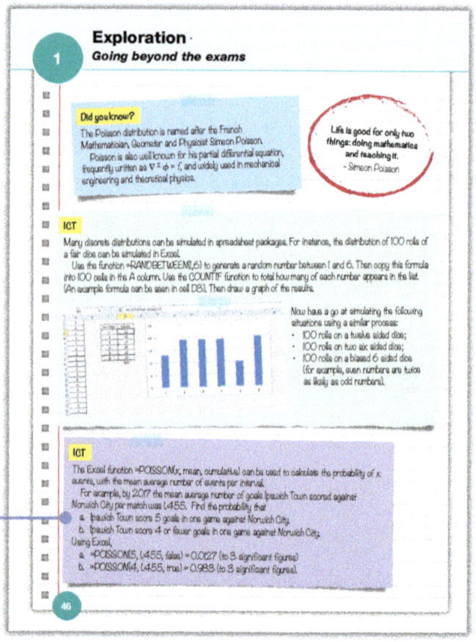

1 Exploration

Going beyond the exams

Did you know?

ICT

Reasoning and problem-solving

Full A Level

Strategy

To solve problems involving type I and type II errors

1. If necessary, decide on the significance level for the test.
2. Calculate the critical value.
3. Calculate the probability of making a type I or type II error.
4. Answer the question in context.

Example 4

A Normal distribution with standard deviation 4 is being tested with hypotheses $H_0: \mu = 7.7$ and $H_1: \mu < 7.7$. Calculate the critical value and hence the probability of making a type II error if the true mean is 2.5 when the significance level is

a 10% b 5% c 1%

a For significance level 10%, the critical value is 2.57

$P(X > 2.57 \mid \mu = 2.5) = 49.3\%$

b The critical value is 1.12

$P(X > 1.12 \mid \mu = 2.5) = 63.4\%$

c The critical value is −1.61

$P(X > -1.61 \mid \mu = 2.5) = 84.8\%$

(2) Find the critical value.

Cumulative probabilities for the Normal distribution can be found using your calculator.

(3) Calculate the probability of a type II error.

Consider a Poisson distribution being tested with hypotheses $H_0: \lambda = 10$ and $H_1: \lambda > 10$. At the 5% significance level the critical region is $X \geq 16$. The table shows the power for a range of true parameter values.

$X \sim Po(\lambda)$

True value of λ	10	11	12	13	14	15	16	17	18
P(*X* **not** in critical region) $= P(X \leq 15)$ $= \beta$	0.951	0.907	0.844	0.764	0.669	0.568	0.466	0.371	0.287
Power of test $= 1 - \beta$	0.049	0.093	0.156	0.236	0.331	0.432	0.534	0.629	0.713

66 Hypothesis testing and association testing Errors in hypothesis testing

Fluency techniques are built on in the **Reasoning and problem-solving** (AO2 and AO3) section.

Strategy boxes help build problem-solving techniques.

Circled numbers show how each step of an example is linked to the strategy box.

'Full A Level' tags show where content is only needed if you are studying for a full A Level in Further Maths. AS students can ignore this content.

The questions in each exercise increase in difficulty. For extra challenge, watch out for the last question or two in each problem-solving (B) exercise.

Exercise 1.3B Reasoning and problem-solving

Full A Level

1 A fair dice is thrown until a one appears for the first time and *X* is a random variable for the number of throws required.

a Write the probability distribution of *X*

b Find the probability that

i $X = 4$ ii $X \geq 3$

c Calculate the mean and variance of *X*

2 A set of twelve cards is numbered 1–12. Cards are chosen at random with replacement until a number greater than 10 is obtained.

a Find the expected value of *X*, a random variable for the total number of cards chosen.

b What is the probability that a number greater than 10 occurs for the first time on the fourth pick?

3 An ordinary dice is thrown until a prime number appears. Find the mean and variance of *X*, the number of throws required.

Assessment sections at the end of each chapter test everything covered within that chapter.

5 The discrete random variable *V* has pdf $f(v) = \begin{cases} \frac{v^2}{30}; x = 1, 2, 3, 4 \\ 0; \text{otherwise} \end{cases}$

a Calculate the median of *V* [2]

b Calculate $E(5V+3)$ and $Var(5V+3)$ [7]

6 The random variable *X*, which takes integer values, is such that $X \sim Po(5.5)$

a Calculate the range of values of *a* such that $P(X < a) > 0.7$ [2]

b Calculate the range of values of *b* such that $P(X > b) < 0.1$ [2]

c Calculate the probability that the value of *X* is within one standard deviation of the mean. [5]

A second, independent, random variable *Y* is such that $Y \sim Po(2)$

d Calculate $P(X + Y < 9)$ [2]

$\frac{1}{A}; t = 1, 2, 3$

1 Assessment

1 The probability distributions of *X* and *Y* are as follows

$P(X = x) = \begin{cases} \frac{x}{k}; x = 1, 2, 3, 4 \\ 0; \text{otherwise} \end{cases}$

y	1	2	3	4
P(Y=y)	0.3	a	a	b

Given that $P(X = 4) = P(Y = 4)$

a Find the value of *k*, *a* and *b* [5]

b Which discrete random variable has the highest value on average?

Explain how you know [4]

1 Discrete random variables

Drivers are required by law to be insured against accidents because the outcome can be expensive for themselves and for other people involved. To calculate the insurance payments to be made by an individual driver, many factors that relate to their circumstances are taken into account. For example, their age, how many years they have been driving, where they live and the age and type of car they will be driving are considered. These factors all have an estimated probability of having an accident associated with them and these probabilities are used to calculate the cost of premiums to drivers.

This is just one area where actuaries use expected values to inform the likelihood of events occurring and to calculate insurance payments for travel, homes against theft, property against fire and health of both people and animals. The associated mathematics is used extensively throughout industry and business.

Orientation

What you need to know

Maths Ch9

- Calculate the mean.
- Calculate the variance.

Maths Ch10

- To model a probability distribution using the binomial distribution.

What you will learn

- To calculate expectation values for given random variables.
- To use Poisson probability models.
- To use the geometric and negative binomial probability models. A Level
- To apply the Central Limit Theorem.
- To derive probability generating functions for given random variables and use them to calculate summary statistics.

What this leads to

Ch 3 Continuous random variables

- Calculating the mean of a continuous random variable.
- Calculating the variance of a continuous random variable.

1.1 Discrete distributions and expectation

Fluency and skills

The probability distribution for a random experiment shows how the total probability of 1 is distributed between all the possible outcomes.

The table shows the probability distribution for X, the number of heads obtained when you throw a fair coin three times.

Number of heads, x	0	1	2	3
$P(X=x)$	$\frac{1}{8}$	$\frac{3}{8}$	$\frac{3}{8}$	$\frac{1}{8}$

You can define a distribution using a probability distribution function. In this case,

$$P(X=x)={}^3C_x\left(\frac{1}{2}\right)^x\left(\frac{1}{2}\right)^{3-x}=\left(\frac{1}{8}\right){}^3C_x;\ x=0,1,2,3$$

See Maths Ch10.2
For a reminder of the binomial distribution.

As with all distributions, you can define measures of location and spread.

In the experiment described you throw a fair coin three times and count the number of heads obtained, X. If you repeat this experiment a large number of times, you will get a large sample of values from the distribution of X. The mean of this sample will give you an estimate of the mean of the distribution of X, and the bigger the sample is, the more accurate the estimate will be.

As the number of experiments increases, the **relative frequency** of each result approaches the **theoretical probability** of that result, so over a series of 800 experiments you expect X to take the value two about 300 times, because $P(X=2)=\frac{3}{8}=\frac{300}{800}$

Number of heads, x	0	1	2	3
$P(X=x)$	$\frac{1}{8}$	$\frac{3}{8}$	$\frac{3}{8}$	$\frac{1}{8}$
Expected frequency in 800 throws	$800\times\frac{1}{8}=100$	$800\times\frac{3}{8}=300$	$800\times\frac{3}{8}=300$	$800\times\frac{1}{8}=100$

In this *expected* sample of 800 values, the sample mean would be

$$\frac{0\times100+1\times300+2\times300+3\times100}{800}=0\times\frac{1}{8}+1\times\frac{3}{8}+2\times\frac{3}{8}+3\times\frac{1}{8}=\frac{3}{2}$$

Over a large number of experiments the mean value of X is expected to be $\frac{3}{2}$. This is an estimate of the population mean for X, and you say the **expected value** of X, $E(X)$, is $\frac{3}{2}$

The **expected value** of a random variable is the number of occurrences of the event that you would expect to get if the relative frequency equals the probability.

The expected value of X will not always be one of the possible values for X, and it will not always be an integer.

This illustrates a general result for discrete random variables.

Key point

The expected value or mean value of a discrete random variable is

$$E[X]=\sum_{\forall x} xP(X=x)$$

The $\forall$ symbol means 'for all' so $\forall x$ means 'for all x.'

Since this is a mean, it is sometimes given the symbol μ, a Greek m.

You will sometimes need to find the expected value of a function of a random variable.

Key point

The **expected value** of a function, $g(X)$, of a discrete random variable X is given by

$$E[g(X)]=\sum_{\forall x} g(x)P(X=x)$$

For example, for a random variable X, you could work out the expected value of X^2 by multiplying each value of x^2 by the corresponding probability and summing the results.

Example 1

A random variable, X, has the probability distribution shown in the table.

x	−1	2	3	5
$P(X=x)$	0.2	0.3	0.4	0.1

Find the expected value of **a** X **b** X^2

a $E[X]=\sum_{\forall x} xP(X=x)$

$=-1\times0.2+2\times0.3+3\times0.4+5\times0.1$

$=2.1$

b

x	−1	2	3	5
x^2	1	4	9	25
$P(X=x)$	0.2	0.3	0.4	0.1

Work out the values of x^2

$E[X^2]=\sum_{\forall x} x^2P(X=x)$

$=1\times0.2+4\times0.3+9\times0.4+25\times0.1$

$=7.5$

The variance of a distribution is given by $\sigma^2=\dfrac{\sum_{\forall x}(x-\mu)^2}{n}$ where $\mu=E(X)$ is the mean. This is the 'expected value of the squared deviations from the mean.' As the variance is the expectation of a function of x, you can use the formula for $E[g(x)]$

See Maths Ch9.2 For a reminder of the variance.

The variance of a random variable X is given by

$$\text{Var}(X) = E[(X-\mu)^2]$$
$$= \sum_{\forall x}(x-\mu)^2 P(X=x)$$

μ can be written as E(X) so this can also be written

$$\text{Var}(X) = \text{E}[(X-\text{E}(X))^2]$$
$$= \sum_{\forall x}(x-\text{E}(X))^2 P(X=x)$$

To make calculations simpler, you usually use an alternative form for variance.

$$\text{Var}(X) = \text{E}(X^2) - \mu^2$$
$$= \text{E}(X^2) - [\text{E}(X)]^2$$
$$= \sum_{\forall x} x^2 P(X=x) - [\text{E}(X)]^2$$

You can write the variance of X as Var(X) or σ^2

The square root of the variance, σ, is the **standard deviation** of the distribution.

Key point

$$\text{Var}(X) = \text{E}(X^2) - [\text{E}(X)]^2$$

Example 2

A random variable, X, has the probability distribution shown in the table.

Find the expected value and variance of X.

x	1	2	3	4
$P(X=x)$	0.3	0.2	0.4	0.1

$$\text{E}(X) = \sum_{\forall x} xP(X=x)$$
$$= 1\times0.3 + 2\times0.2 + 3\times0.4 + 4\times0.1$$
$$= 2.3$$

x	1	2	3	4
x^2	1	4	9	16
$P(X=x)$	0.3	0.2	0.4	0.1

Work out the possible values of x^2

$$\text{Var}(X) = \text{E}(X^2) - [\text{E}(X)]^2$$
$$= 1\times0.3 + 4\times0.2 + 9\times0.4 + 16\times0.1 - 2.3^2$$
$$= 1.01$$

Exercise 1.1A Fluency and skills

1 Calculate the expectation and variance for each of these variables.

a

x	1	4	9	16	25	36
$P(X=x)$	$\frac{1}{6}$	$\frac{1}{6}$	$\frac{1}{6}$	$\frac{1}{6}$	$\frac{1}{6}$	$\frac{1}{6}$

b

y	−1	0	1	2
$P(Y=y)$	$2a$	$2a$	a	a

2 For each of these probability distributions, find the mean and the variance.

a $P(X=a) = \frac{1}{4};\ a = 1, 2, 3, 4$

b $P(X=b) = \frac{b}{3};\ b = 1, 2$

c $P(X=c) = \begin{cases} \frac{1}{c}; c = 2, 3 \\ \frac{1}{6}; c = 4 \end{cases}$

3 A fair ten-sided dice with faces numbered 1–10 is thrown once. Prove that the expectation and variance of the score obtained are $\frac{11}{2}$ and $\frac{33}{4}$

4 Five cards are numbered from 1 to 5. Two cards are drawn at random with replacement and the total, T, noted. Find the expected value of T.

5 A four-sided spinner with sides numbered 1–4 is biased so that odd-numbered outcomes are each twice as likely as even-numbered outcomes. Find the expected value and variance of the score on a randomly chosen spin.

6 A spinner can land on numbers 1–9 with equal probability. Show that the expected score for one throw is 5 and find the variance of this score.

7 A fair dice is thrown once. Find the mean value of the **square** of the score.

8 A six-sided dice with faces numbered 1–6 is biased so that the probability of any score is proportional to the square of the score. Find the expected value of S, the score on a randomly chosen throw.

9 A fair dice is thrown once. If R is the reciprocal of the score, find the mean value of R.

10 X is a random variable for the score on a biased tetrahedral dice. Its probability distribution is shown in the table. Find its mean and variance.

x	1	2	3	4
$P(X=x)$	0.4	0.2	0.2	0.2

11 A number is chosen at random from the integers 1 to 30

a Find the expected number of prime numbers in this sample of size 1 (note that your answer will be a fraction).

b Explain what this means for a set of repeated trials of this experiment.

12 A fair coin is tossed until a head or four tails appear.

a Give the probability distribution of X, the number of throws required.

b What is the expected number of throws?

Reasoning and problem-solving

Strategy

To solve a problem involving expectation

1. Find the probability distribution of the random variable, either in the form of a function or as a table.
2. Use standard formulae to find the mean and variance of the random variable.
3. Where necessary, find the expectation value of a function of X, g(X)
4. Give your conclusion.

Let X be a binomial random variable, $X \sim \mathrm{B}(n, p)$. There are n trials, each with the two possible outcomes: success or failure. X is the number of successes in n trials, where the probability of success is constant at p.

You can write X as the sum of Y_i, $i = 1, \ldots, n$ where Y_1 is the number of successes in the first trial (0 or 1), Y_2 is the number of successes in the second trial, and so on.

$X = Y_1 + Y_2 + \ldots + Y_n$; Y_i = number of successes in the ith trial

Y_i are identical independent **Bernoulli variables**; this type of variable can only take the values 0 and 1, the probability of 1 is p and the probability of 0 is $(1 - p)$

$$\mathrm{E}(Y_i) = \sum_{\forall y} y\mathrm{P}(Y = y) = 0 \times (1-p) + 1 \times p = p$$

$$\begin{aligned}\mathrm{Var}(Y_i) &= \sum_{\forall y} y^2 P(Y_i = y) - [E(Y_i)]^2 \\ &= 0^2 \times (1-p) + 1^2 \times p - [p]^2 \\ &= p - p^2 \\ &= p(1-p)\end{aligned}$$

The variables Y_i have $\mathrm{E}(Y_i) = p$ and $\mathrm{Var}(Y_i) = p(1 - p)$

Because the variables Y_i are independent we can use the results:

For two random variables A and B, $\mathrm{E}(A + B) = \mathrm{E}(A) + \mathrm{E}(B)$

For two independent random variables, A and B,
$\mathrm{Var}(A + B) = \mathrm{Var}(A) + \mathrm{Var}(B)$

The proof of these results is not required in this course.

$$\begin{aligned}\mathrm{E}(X) &= \mathrm{E}(Y_1) + \mathrm{E}(Y_2) + \ldots + \mathrm{E}(Y_n) \\ &= p + p + \ldots + p \\ &= np\end{aligned}$$

$$\begin{aligned}\mathrm{Var}(X) &= \mathrm{Var}(Y_1) + \mathrm{Var}(Y_2) + \ldots + \mathrm{Var}(Y_n) \\ &= np(1 - p)\end{aligned}$$

Key point

For $X \sim \mathrm{B}(n, p)$, $\mathrm{E}(X) = np$ and $\mathrm{Var}(X) = np(1 - p)$

Example 3

The discrete random variable X follows a binomial distribution $X \sim \mathrm{B}(25, 0.3)$

a Find the mean and variance of X

b Show that less than 1% of values lie more than three standard deviations away from the mean.

a $\mathrm{E}(X) = np = 25 \times 0.3 = 7.5$

$\mathrm{Var}(X) = np(1 - p) = 25 \times 0.3 \times (1 - 0.3) = 5.25$

Use the formulae for mean and variance of a binomial distribution.

b $7.5 + 3 \times \sqrt{5.25} = 14.3\ldots$ $\quad 7.5 - 3 \times \sqrt{5.25} = 0.626\ldots$

$P(X > 14.3) = 1 - P(X \le 14) = 1 - 0.99822\ldots = 0.0017784\ldots$

Use the binomial distribution function on your calculator.

$P(X < 0.626) = P(X = 0) = 0.0001341\ldots$

X takes integer values.

P(X lies more than three standard deviations away from the mean)
$= 0.0017784\ldots + 0.0001341\ldots = 0.001912\ldots < 0.01$

You will sometimes need to calculate the expectation of functions of random variables.

Example 4

An ordinary six-sided dice is thrown once. The score obtained, X, is a random variable. The random variable Y is defined by the equation $Y = 5X - 3$.

By finding the possible values of Y, show that

a $E(Y) = 5E(X) - 3$ **b** $Var(Y) = 25Var(X)$

a Since X takes values 1 – 6, Y takes values 2, 7, 12, 17, 22, 27

x	1	2	3	4	5	6
y	2	7	12	17	22	27
p	$\frac{1}{6}$	$\frac{1}{6}$	$\frac{1}{6}$	$\frac{1}{6}$	$\frac{1}{6}$	$\frac{1}{6}$

① Find the probability distribution of the random variable in the form of a table.

$$E(X) = 1\times\frac{1}{6} + 2\times\frac{1}{6} + 3\times\frac{1}{6} + 4\times\frac{1}{6} + 5\times\frac{1}{6} + 6\times\frac{1}{6}$$
$$= 3.5$$

② Use standard formulae to find the mean of the random variable.

$$5E(X) - 3 = 5\times 3.5 - 3$$
$$= 14.5$$

$$E(Y) = 2\times\frac{1}{6} + 7\times\frac{1}{6} + 12\times\frac{1}{6} + 17\times\frac{1}{6} + 22\times\frac{1}{6} + 27\times\frac{1}{6}$$
$$= 14.5$$

Therefore $E(Y) = 5E(X) - 3$ as required.

④ Give your conclusion.

b $Var(Y) = E(Y^2) - [E(Y)]^2$

$$= 2^2\times\frac{1}{6} + 7^2\times\frac{1}{6} + 12^2\times\frac{1}{6} + 17^2\times\frac{1}{6} + 22^2\times\frac{1}{6} + 27^2\times\frac{1}{6} - 14.5^2$$
$$= \frac{875}{12}$$

② Use standard formulae to find the variance of the random variable.

$Var(X) = E(X^2) - [E(X)]^2$

$$= 1^2\times\frac{1}{6} + 2^2\times\frac{1}{6} + 3^2\times\frac{1}{6} + 4^2\times\frac{1}{6} + 5^2\times\frac{1}{6} + 6^2\times\frac{1}{6} - 3.5^2$$
$$= \frac{35}{12}$$

$$\frac{875}{12} = 25\times\frac{35}{12}$$

Therefore $Var(Y) = 25Var(X)$ as required.

④ Give your conclusion.

Example 4 illustrates a general result.

Key point

If X and Y are two discrete random variables and $Y = aX + b$, where a and b are constants, then

$E(Y) = aE(X) + b$ and $Var(Y) = a^2Var(X)$

You can also write this as $E(aX + b) = aE(X) + b$ and $Var(aX + b) = a^2Var(X)$

See Maths 9.2
For a reminder of coding.

Example 5

A random variable, S, has the probability distribution shown in the table, where a is a positive constant.

s	1	2	3
$P(S=s)$	$\frac{1}{2}$	a	$2a$

a Find the value of a

The variable T is related to S by the equation $T=-3S+6$

b Find the mean and variance of S and hence the mean and variance of T.

a $3a+\frac{1}{2}=1 \rightarrow a=\frac{1}{6}$ — Find the probability distribution. **1**

b $E(S)=1\times\frac{1}{2}+2\times\frac{1}{6}+3\times\frac{1}{3}=\frac{11}{6}$ — Use standard formulae to find the mean of the random variable. **2**

$Var(S)=1^2\times\frac{1}{2}+2^2\times\frac{1}{6}+3^2\times\frac{1}{3}-\left(\frac{11}{6}\right)^2=\frac{29}{36}$ — Use standard formulae to find the variance of the random variable. **2**

$T=-3S+6$

$E(T)=-3E(S)+6=-3\times\frac{11}{6}+6=\frac{1}{2}$ — Use $E(Y)=aE(X)+b$

$Var(T)=(-3)^2Var(S)=9\times\frac{29}{36}=\frac{29}{4}$ — Use $Var(aX+b)=a^2Var(X)$

Exercise 1.1B Reasoning and problem-solving

1 A random variable X has the probability distribution shown in the table.

x	1	2	3	4	5	6
$P(X=x)$	$\frac{1}{6}$	$\frac{1}{6}$	$\frac{1}{6}$	$\frac{1}{6}$	a	b

Given that the mean of X is $\frac{52}{15}$, find the values of a and b

2 R is a random variable with the given distribution.

r	-3	-2	-1	0	1	2	3
$P(R=r)$	$\frac{1}{7}$	$\frac{1}{7}$	$\frac{1}{7}$	$\frac{1}{7}$	$\frac{1}{7}$	$\frac{1}{7}$	$\frac{1}{7}$

A single value of R is chosen at random.

a Find the mean and standard deviation of R

b What is the probability that the value chosen will be one standard deviation or closer to the mean?

3 The proportion of boys born in the UK is 0.51 of all live births. Four UK births are selected at random.

a Write down the probability distribution for X, a random variable for the number of girls born in the four births.

b Find the mean and variance of X

4 $X \sim B(n, p)$. $E(X) = 8.1$ and $Var(X) = 4.455$, find the values of n and p

5 Jenny and John play a dice game. John says that he will get at least one six in four throws of a fair dice and will pay Jenny £1 if he fails.

a What is the probability of John winning any given game?

b How much should Jenny pay to enter the game so that in the long run neither player can expect to win anything? Give your answer to the nearest 1p.

6 A bag contains one red and two black balls. A ball is drawn from the bag and then, without the first ball being replaced, a second ball is drawn. By first finding the probability distribution, find the expected value of B, the number of black balls drawn.

7 A random variable X has the probability distribution function

$$P(X = x) = \begin{cases} \frac{1}{5n}; x = 1,2,3,...,5n \\ 0; \text{otherwise} \end{cases}$$

where n is a positive integer.

You may assume that

$$\sum_{i=1}^{n} i = \frac{n(n+1)}{2} \quad \text{and that} \quad \sum_{i=1}^{n} i^2 = \frac{n}{6}(n+1)(2n+1)$$

a Prove that $E[X] = \frac{(1+5n)}{2}$

b Find in terms of n an expression for the variance of X

8 A circle of radius r is inscribed in a square of side $2r$. Three darts are randomly dropped onto the square. D is a random variable for the number of darts that land within the circle.

a Find $E[D]$. (Assume that the probability of a dart landing in the circle is equal to (area of circle ÷ area of square).)

b One dart is dropped randomly onto the square 100 times and on 71 occasions the dart lands within the circle. Use this data to estimate the value of π

1.2 The Poisson distribution

Fluency and skills

You can use the Poisson distribution to model the occurrence of random events in time or space.

A random variable, X, might be modelled by a Poisson distribution with parameter λ, $X \sim \text{Po}(\lambda)$, if

- Events occur at random,
- Events occur independently,
- Events cannot occur simultaneously,
- The probability of an event is constant over time or space.

Key point

If X is the number of random events that occur in a given interval of time or space, λ is the mean number of occurrences in that interval and X follows a Poisson distribution, $X \sim \text{Po}(\lambda)$, then its probability distribution function is given by

$$P(X=x)=\frac{e^{-\lambda}\lambda^x}{x!};\ x=0, 1, 2, \ldots$$

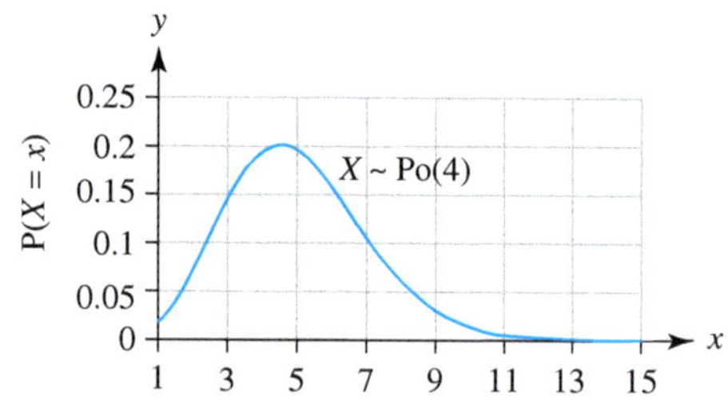

▲ Graph of $P(X=x)$ against x where $X \sim \text{Po}(\lambda)$ and $\lambda = 4$

Some examples of random variables that may follow a Poisson distribution are the number of radioactive disintegrations in five-second intervals, the number of currants in a bun whose dough was well mixed, and the number of telephone calls you receive in one-hour intervals of time.

Random events in continuous time or space constitute a **Poisson process**.

Example 1

A Geiger counter detects radioactive disintegrations at a mean rate of 23 per minute. Use the Poisson distribution function with $\lambda = 23$ to find the probability that, in a randomly chosen minute, there will be

a 20 disintegrations,

b Between 21 and 23 inclusive disintegrations.

a Let X be the number of disintegrations in a randomly chosen minute.

$$P(X=20)=\frac{e^{-23}23^{20}}{20!}$$

$=0.0724$ (3 sf)

Substitute $\lambda = 23$ and $x = 20$ into $P(X=x)=\frac{e^{-\lambda}\lambda^x}{x!}$

b $$P(21 \le X \le 23)=\sum_{x=21}^{23}\frac{e^{-23}23^x}{x!}$$

$$=\frac{e^{-23}23^{21}}{21!}+\frac{e^{-23}23^{22}}{22!}+\frac{e^{-23}23^{23}}{23!}$$

$=0.079+0.083+0.083$

$=0.245$ (3 sf)

Substitute $x = 21$, $x = 22$ and $x = 23$, into $P(X=x)=\frac{e^{-23}23^x}{x!}$

Use the sum function on your calculator to check your answers.

Calculator

Try it on your calculator

You can use a calculator to find probabilities for a Poisson distribution.

```
Poisson      C.D
Data         :Variable
x            :5
μ            :3
Save Res     :None
Execute

None LIST
```

```
Poisson      C.D
    p=0.91608205
```

Activity

Find out how to calculate $P(X \le 5)$ where $X \sim Po(3)$ on *your* calculator.

Example 2

A random variable X has a Poisson distribution with parameter $\lambda = 2$

a Use statistical tables to find $P(X \le 5)$

b Use the probability distribution function of X to copy and complete this table.

x	0	1	2	3	4	5	6	>6
$P(X=x)$	0.135	0.271					0.0120	0.00453

c Use the values $x = 0, 1, 2, \ldots, 6$ to estimate the mean and variance of X

a $\lambda = 2, x = 5$

$P(X \le 5) = 0.9834$

Look at the $\lambda = 2.0$ column and the $x = 5$ row.

b

x	0	1	2	3	4	5	6	>6
$P(X=x)$	0.135	0.271	0.271	0.180	0.0902	0.0361	0.0120	0.00453

$P(X=3) = \frac{e^{-2}2^3}{3!} = 0.180$

c $E(X) \simeq \sum_{x=0}^{6} xP(X=x)$

$= 0 \times 0.135 + 1 \times 0.271 + 2 \times 0.271 + 3 \times 0.180$
$+ 4 \times 0.0902 + 5 \times 0.0361 + 6 \times 0.0120$

$= 1.9663$

$Var(X) \simeq \sum_{x=0}^{6} x^2P(X=x) - [E(X)]^2$

$= 0^2 \times 0.135 + 1^2 \times 0.271 + 2^2 \times 0.271 + 3^2 \times 0.180 + 4^2 \times 0.0902 + 5^2 \times 0.0361$
$+ 6^2 \times 0.0120 - 1.9663^2$

$= 1.886$

You can use the Poisson distribution function to calculate exact values of the mean and variance of the distribution.

If X has a probability distribution function given by

$$P(X=x) = \frac{e^{-\lambda}\lambda^x}{x!};\ x = 0, 1, 2, \ldots$$

then the mean and variance of X are respectively given by

$$E(X) = \sum_{x=0}^{\infty} xP(X=x) = \lambda \qquad Var(X) = \sum_{x=0}^{\infty} x^2P(X=x) - \lambda^2 = \lambda$$

Key point

For $X \sim Po(\lambda)$, $E(X) = \lambda$ and $Var(X) = \lambda$

Example 3

Compare your answers to part **a** with the estimates found in Example 2.

a Write the mean and variance of a random variable X with a Poisson distribution, $\lambda = 2$

b Find the probability that a randomly chosen value of X will be within 1 standard deviation of the mean.

a $X \sim Po(\lambda = 2)$

mean = 2, variance = 2

b Standard deviation $\sqrt{2} = 1.414$ (3 dp)

$P(2 - 1.414 < X < 2 + 1.414) = P(0.586 < X < 3.414)$

$= P(X = 1 \text{ or } 2 \text{ or } 3)$

$= 0.271 + 0.271 + 0.180$

$= 0.722$

Exercise 1.2A Fluency and skills

1 A random variable X has a Poisson distribution, mean 4. Find the probability that X takes values 0, 1, 2, 3, > 3

2 Random events occur at a rate of three per ten seconds. Find the probability that there will be at least two events in any randomly chosen ten-second interval.

3 A random variable R is Poisson-distributed, parameter 4. Find the probability that R takes values 0 or greater than 4

4 A random variable X has a Poisson distribution with parameter $\lambda = 1.1$

a Using the probability distribution function of X, copy and complete the table.

x	0	1	2	3	4	>4
$P(X = x)$	0.333	0.366				

b Sketch the probability distribution function of X for $x = 0, 1, 2, 3, 4$

5 Random events occur at an average rate of 10 per 12 minutes. Find the probability that there will be fewer than three events in any randomly chosen 12-minute interval.

6 Radioactive decays occur at an average rate of 34 per 12-second interval. Find the probability that in two randomly chosen 12-second intervals, there will be at least 30 disintegrations in exactly one of the intervals.

7 A household receives on average four telephone calls per day. What is the probability that in a randomly chosen week (of seven days), there is at least one call every day? State any assumptions you make about the appropriate probability distribution.

8 Emails arrive at your account randomly at a rate of four per two hours. Find the probability that in a randomly chosen two hours, you will receive at least three emails.

9 There are on average four hazelnuts in a 200 g chocolate bar. What is the probability that in five randomly chosen bars, there is at least one hazelnut in each bar? State any distributional or other assumptions you make.

10 A fabric-weaving machine produces cloth with flaws, which occur at a rate of 3.4 per square metre. A customer orders a 2 m^2 piece of cloth and insists that it must be flawless. Find the probability that the first piece of cloth produced for the customer meets her requirements.

11 Cars pass a traffic checkpoint at a rate of two per hour. Find the probability that in two consecutive hours, there is a total of four cars passing. You should treat the two one-hour intervals separately.

Reasoning and problem-solving

Strategy

To solve a problem involving the Poisson distribution

1. Ensure that the conditions for a Poisson distribution apply to the problem.
2. Identify the value of λ and write the distribution function.
3. Use a calculator, statistical tables or the Poisson distribution function to find probabilities.
4. Where required, use standard results to find the mean and variance of the distribution.
5. Give your conclusion.

In n equal intervals of time or space, the total number of random events occurring equals the sum of the number of events in each of the separate intervals.

Key point

Let X_i; $i = 1, 2, ..., n$ be a set of n independent and identically distributed random variables each with a Poisson distribution, mean λ. If $X_T = \sum_{i=1}^{n} X_i$, then X_T has a Poisson distribution with mean $n\lambda$

Example 4

Events occur randomly at a mean rate of λ per hour.

a By considering the two hours individually find, in terms of λ, the probability that three events occur in a randomly chosen two-hour period.

b Show that the number of events in a randomly chosen two-hour period follows a Poisson distribution, and give the mean value.

c If $\lambda = 5$, find the probability that fewer than 14 events occur in a three hour period.

a Let X_1 = number of events in the first hour, X_2 = number of events in the second hour.

$X_T = X_1 + X_2$

$P(X_T = 3) = P(X_1 = 3, X_2 = 0 \text{ or } X_1 = 2, X_2 = 1 \text{ or } X_1 = 1, X_2 = 2 \text{ or } X_1 = 0, X_2 = 3)$

$$P(X_T = 3) = \frac{e^{-\lambda}\lambda^3}{3!} \times \frac{e^{-\lambda}\lambda^0}{0!} + \frac{e^{-\lambda}\lambda^2}{2!} \times \frac{e^{-\lambda}\lambda^1}{1!} + \frac{e^{-\lambda}\lambda^1}{1!} \times \frac{e^{-\lambda}\lambda^2}{2!} + \frac{e^{-\lambda}\lambda^0}{0!} \times \frac{e^{-\lambda}\lambda^3}{3!}$$

$$= \frac{e^{-2\lambda}\lambda^3}{3!} + \frac{e^{-2\lambda}\lambda^3}{2!} + \frac{e^{-2\lambda}\lambda^3}{2!} + \frac{e^{-2\lambda}\lambda^3}{3!}$$

X_T = total number of events in the two hours.

A total of 3 events in two hours means 3 in the first and 0 in the second or 2 in the first and 1 in the second...

2 Use the probability distribution function to define the required probabilities.

(Continued on next page)

$$= e^{-2\lambda}\lambda^3\left(\frac{1}{3!}+\frac{1}{2!}+\frac{1}{2!}+\frac{1}{3!}\right)$$

$$= \frac{4}{3}e^{-2\lambda}\lambda^3$$

b $\frac{4}{3}e^{-2\lambda}\lambda^3 = e^{-2\lambda}\lambda^3 \times \frac{8}{6} = \frac{e^{-2\lambda}(2\lambda)^3}{3!}$

This is the probability distribution function of a Poisson distribution, mean 2λ

c Let Y = number of events per three-hour period.

$Y \sim Po(15)$

$P(Y < 14) = P(Y \le 13)$

$= 0.363$

② The parameter is $3 \times \lambda = 3 \times 5$

③ Use the Poisson function on your calculator.

The summative property of Poisson random variables also applies to variables with different means.

If X and Y are independent Poisson random variables with parameters λ and μ then the variable $X+Y$ follows a Poisson distribution with parameter $\lambda+\mu$

Key point

If X and Y are independent random variables such that $X \sim \text{Po}(\lambda)$ and $Y \sim \text{Po}(\mu)$ then $X+Y \sim \text{Po}(\lambda+\mu)$

Example 5

Two unrelated companies share a floor in an office block. The companies receive, on average, X phone calls and Y phone calls respectively in an hour. $X \sim \text{Po}(15)$ and $Y \sim \text{Po}(27)$. The times in this question all occur during regular working hours.

a Calculate the probability that, in a given hour, 30 phone calls are received on the floor.

b Calculate the probability that more than 300 calls are received on the floor in a given seven-hour period.

a $Z = X + Y$ is the number of phone calls received on the floor in an hour.

$Z \sim Po(42)$

$P(Z = 30) = \frac{42^{30}e^{-42}}{30!} = 0.0108$

② The parameter is $15 + 27 = 42$

b Let Z_7 be the number of calls on the floor in a seven-hour period.

Mean $= 7 \times 42 = 294$

② Mean $= n\lambda$

$Z_7 \sim Po(294)$

$P(Z_7 > 300) = 1 - P(Z_7 \le 300) = 1 - 0.6507\ldots = 0.349$

Under certain conditions you can use the Poisson distribution to provide a good approximation to the binomial distribution.

If the random variable X has a binomial distribution $X \sim B(n,p)$, then as n becomes very large and p very small, the distribution of X tends to that of a Poisson random variable with parameter np

Key point

If $X \sim B(n,p)$ then for large n and small p, $X \sim Po(np)$

In practice, the larger n and the smaller p, the better the approximation is.

Example 6

While learning to touch type, a student makes on average two mistakes every 17 characters. For an exam, she types a paragraph of 71 letters.

a Write the exact distribution of X, the number of errors in 71 letters, stating any assumptions you make.

b State the mean of a Poisson distribution which can be used to approximate the distribution in part **a**.

The student is awarded a distinction if she makes at most three mistakes in the passage.

c Calculate the approximate probability that she will be awarded a distinction using

i The distribution found in part **a**,

ii The approximating Poisson distribution found in part **b**.

a $X \sim B\left(71, \frac{2}{17}\right)$

Assume that the probability of a mistake in any letter is a constant and the outcome for each letter (error or no error) is independent of that for all other letters.

b $X \sim Po(8.353)$

Use the binomial mean, $np = 71 \times \frac{2}{17}$

c i $X \sim B\left(71, \frac{2}{17}\right)$

P(distinction) $= P(X \le 3) = 0.0263$

ii $X \sim Po(8.353)$

P(distinction) $= P(X \le 3) = 0.0333$ (3 sf)

3 Use a calculator or statistical tables to find the required probabilities.

Exercise 1.2B Reasoning and problem-solving

1 Random events occur at a rate of three per ten seconds. Find the probability that there will be at most 33 events in any randomly chosen two-minute interval.

2 Cars pass a traffic checkpoint at a rate of two per minute. Find the probability that, in a randomly selected two-minute interval, a total of four cars pass.

3 Shoppers arrive at a supermarket at an average rate of seven every minute. Find the probability that in a randomly chosen one-hour interval at least 400 shoppers will arrive.

4 At a certain set of traffic lights there is on average one road traffic accident (RTA) every nine days. Find the probability that in a randomly chosen month (of 31 days) there is at most one accident.

5 $X \sim B(60, 0.02)$

a Say why this distribution can be approximated by a Poisson distribution, and state the approximating Poisson distribution.

b Use the Poisson approximation to calculate

i $P(X=2)$ ii $P(X>2)$ iii $P(3 \leq X \leq 6)$

6 $X \sim B(75, 0.01)$. State the mean and variance of X

a Using the Binomial distribution,

b Using a Poisson approximation to the Binomial distribution.

7 I receive on average 14 emails per day. Assuming that their arrival constitutes a random process, what is the probability that in a randomly chosen seven days

a I receive at least twelve emails every day

b I receive a total of at least 84 emails?

8 Two radioactive substances, A and B, emit particles independently and the emissions follow Poisson distributions with means 3.2 and 4.15 particles per minute respectively. Find the probability that overall, in a given minute, there will be

a 7 emissions,

b More than 7 emissions,

c 2 emissions from one substance and 5 emissions from the other.

9 Drops of red and blue paint are scattered over tiles such that the number of drops of red paint, R, follows a Poisson distribution with mean 1.4 drops per cm^2 and the number of drops of blue paint, B, follows a Poisson distribution with mean 2.7 drops per cm^2. R and B are independent.

Find the following probabilities.

a A randomly chosen 1 cm^2 area of tile contains between 3 and 5 drops of paint, inclusive.

b A square tile with side length 10 cm contains more than 400 drops of paint.

c In a sample of 20 square tiles with side length 10 cm, no more than 15 tiles contain more than 400 drops of paint.

10 At a weather station during light rain, raindrops land on a one-square-metre detector plate at a rate of 25 per minute. Assuming that their arrival constitutes a random process, what is the probability that in a randomly chosen ten-minute period there will be no rain detected on a $10\,cm \times 20\,cm$ plate? State any assumptions you make.

11 Silver chain is made by linking circles of silver. Quality management rules require that the first three 5 m lengths of chain produced in a day have at most one flaw each, otherwise the machine must be adjusted. Find the probability that no adjustment is necessary if imperfections in the chain occur at a rate of 3.2 per 10 metres.

12 A carpet-weaving machine produces carpet with an average of two flaws per $5\,m^2$. Assuming that the flaws are randomly distributed, find the probability that a randomly chosen area of $12.5\,m^2$ contains

a At most three flaws. b More than ten flaws.

1.3 The geometric distribution

Full A Level section

Fluency and skills

Full A Level

Given a sequence of independent trials, each with the same probability, p, of success, let the discrete random variable, X, represent the number of trials required to achieve the first success. X follows a geometric distribution with parameter p, the probability of success in a given trial.

The geometric distribution can be used to model discrete random variables such as the number of throws of a dice up to the first six or the number of births in a hospital on a particular day up to the first boy being born. So in general, it models the number of independent trials (each with probability p of success) until a success occurs.

Key point

If X is a geometric random variable with parameter p, $X \sim \text{Geo}(p)$, then its probability distribution function is given by

$$P(X = x) = (1 - p)^{x-1}p;\ x = 1, 2, \ldots$$

The density function is positively skewed and is defined for positive integer values of x. The graph shows the function for $p = 0.5$

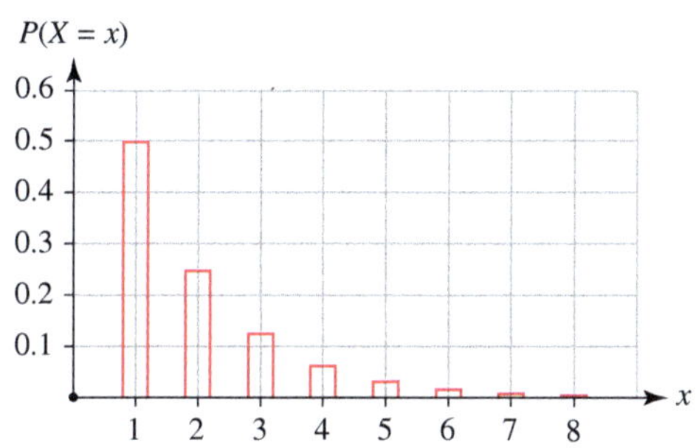

Example 1

$X \sim \text{Geo}(0.72)$

Find

a $P(X = 5)$ **b** $P(X \le 5)$ **c** $P(X > 5)$

a $P(X = 5) = (1 - 0.72)^{5-1}(0.72) = 0.00443$ (3sf)

b $P(X \le 5) = \sum_{x=1}^{5}(1 - 0.72)^{x-1}(0.72) = 0.998\ldots$

$= 0.998$ (3sf)

Check your answer using the geometric distribution function on your calculator.

c $P(X > 5) = 1 - P(X \le 5) = 1 - 0.998\ldots = 0.00172$ (3sf)

It is possible to find the mean and variance of a geometric random variable.

Example 2

$X \sim \text{Geo}\left(\frac{1}{2}\right)$. By considering x-values 1, 2, ..., 8, show that to 1 decimal place $E(X) = \frac{1}{p}$

$X \sim Geo\left(\frac{1}{2}\right) \rightarrow P(X=x) = \left(1-\frac{1}{2}\right)^{x-1}\left(\frac{1}{2}\right) = \left(\frac{1}{2}\right)^x; x = 1, 2, 3, \ldots$

Therefore probabilities for $x = 1, 2, \ldots, 8$ are $\frac{1}{2}, \frac{1}{4}, \frac{1}{8}, \ldots, \frac{1}{256}$

Therefore $E(X) \approx 1\times\frac{1}{2} + 2\times\frac{1}{4} + \ldots + 8\times\frac{1}{256} = 2.0$ (1 dp)

But $\frac{1}{p} = 2$, therefore $E(X) = \frac{1}{p}$ to one decimal place.

This result, along with the equivalent result given for variance, is exact but the proof of this is beyond the scope of this course.

Key point

A geometric random variable, X, with distribution function $P(X=x) = (1-p)^{x-1}p$; $x = 1, 2, 3, \ldots$ has

$\mu = E[X] = \frac{1}{p}$ and $\sigma^2 = \text{Var}[X] = \frac{1-p}{p^2}$

Example 3

A random variable, X, has probability distribution given by $P(X=x) = (1-p)^{x-1}p$; $x = 1, 2, 3, \ldots$
Given that the mean of X is 6, find its variance.

$X \sim Geo(p)$ so $E(X) = \frac{1}{p} = 6$ therefore $p = \frac{1}{6}$

The distribution function shows that X is geometric.

$Var(X) = \frac{1-p}{p^2} = \frac{1-\left(\frac{1}{6}\right)}{\left(\frac{1}{6}\right)^2} = 30$

Exercise 1.3A Fluency and skills

Full A Level

1 A random variable X has a geometric distribution, $p = 0.75$. Find the probability that X takes a value

 a of at least 2 **b** between 1 and 3 inclusive.

2 A random variable G has probability distribution given by $P(G=g) = (1-p)^{g-1}p$; $g = 1, 2, 3, \ldots$
If $p = 0.9$, find

 a $P(G=2)$ **b** $P(G=4)$ **c** $P(G<3)$ **d** $P(G\geq 3)$

3 A random variable X has a geometric distribution with parameter $p = \frac{1}{4}$

 a Using the probability density function of X, copy and complete the table.

x	1	2	3	4	5	6
$P(X=x)$						

 b Sketch the probability distribution function of X for $x = 1, 2, \ldots, 6$

4 a Write the probability distribution function for a random variable X with a geometric distribution, parameter $p = 0.4$

b Show that its cumulative distribution function, defined by $F(a) = P(X \le a)$, can be written $F(a) = 1 - 0.6^a$

c Calculate the mean and variance of X

5 a Find the mean, mode and standard deviation of X, a geometric random variable with parameter $p = 0.25$

b Using one of Pearson's measures of skewness, $\frac{\text{mean} - \text{mode}}{\text{standard deviation}}$, calculate the skewness of X

6 The random variable X has a geometric distribution, parameter p. Let $q = 1 - p$. Prove that

a $P(X < a) = 1 - q^{a-1}$ b $P(X \ge a) = q^{a-1}$

7 The random variable X has a geometric distribution with parameter p

a Write the probability distribution function of X and show that the cumulative distribution function of X, $F(a)$, is $1 - q^a$ where $q = 1 - p$

b Use your answer in part **a** to find the median value of X when $p = \frac{1}{3}$. Use the definition of the median of a discrete random variable as the smallest x such that $P(X \le x) \ge 0.5$

8 Prove that the probability distribution function $P(X = x) = (1 - p)^{x-1}p$; $x = 1, 2, 3, \ldots$ gives probabilities that obey the result $P(S) = 1$ where S is the sample space for the experiment.

Reasoning and problem-solving

Full A Level

Strategy

To solve a problem using the geometric distribution

1. Ensure that the conditions for the geometric distribution apply to the problem.
2. Use a calculator, statistical tables or the geometric distribution function to find probabilities.
3. Where required, use standard results to find the mean and variance of the distribution.

A sequence of independent trials which have two possible outcomes, 'success' or 'failure', and a constant probability of success, p, is known as a sequence of **Bernoulli trials**. Suppose that a set of independent and identical Bernoulli trials was performed, let p be the probability of success in each trial and let X be a random variable for the number of trials up to and including the first success.

The first success on the xth trial means $(x - 1)$ failures followed by a success. Therefore $P(X = x) = (1 - p)^{x-1}p$; $x = 1, 2, \ldots$ This is the distribution function of a geometric distribution.

The number of Bernoulli trials up to but not including the first success also follows a geometric probability distribution – see Exercise 1.3B, Question 6

Key point

The number of Bernoulli trials up to and including the first success follows a geometric probability distribution, mean $\frac{1}{p}$ and variance $\frac{1-p}{p^2}$

Example 4

During May in the north of Spain, the probability of rain occurring on a particular day is $\frac{1}{7}$.

On each day from 1st May, it is noted whether or not it rained. Let X be the number of days up to and including the day it rained for the first time.

a Give the distribution of X. You should state any assumptions you make.

b What is the mean and standard deviation of X?

c What is the probability that in a given year there is no rain during May?

a $X \sim \text{Geometric}\left(\frac{1}{7}\right)$, assuming independence of outcomes from one day to the next.

b Mean $= \frac{1}{p} = 7$; variance $= \frac{1-p}{p^2} = \frac{1-\frac{1}{7}}{\left(\frac{1}{7}\right)^2} = 42$

c $P(X > 31) = \left(\frac{6}{7}\right)^{31} = 0.00841$ (3sf)

1 Since the probability of rain is constant at $\frac{1}{7}$ the conditions for a geometric distribution are ensured.

3 Use standard results to find the mean and variance of the distribution.

2 Use the geometric distribution function to find probabilities.

Example 5

A fair dice is repeatedly thrown until a six occurs.

a Write the distribution of the random variable X, the number of throws up to and including the first six.

b Find the probability that the number of throws up to and including the first six exceeds 10.

c Given that the first 20 throws do not include any sixes, find the probability that the number of **further** throws up to and including the first six exceeds 10.

d Comment on the results in parts **b** and **c**.

a $X \sim \text{Geometric}\left(\frac{1}{6}\right)$

b $P(X > 10) = P(0 \text{ sixes in } 10 \text{ throws}) = \left(\frac{5}{6}\right)^{10} = 0.162$ (3sf)

c P(the number of further throws to the first six exceeds 10|zero sixes in first 20 throws)

$$= P(X > 30 | X > 20)$$

$$= \frac{P(X > 30 \text{ and } X > 20)}{P(X > 20)}$$

$$= \frac{P(X > 30)}{P(X > 20)}$$

$$= \left(\frac{5}{6}\right)^{30} \div \left(\frac{5}{6}\right)^{20} = \left(\frac{5}{6}\right)^{10} = 0.162 \text{ (3 sf)}$$

d The answers are equal. The outcome of the first 20 throws is not 'remembered' in future throws.

1 Since the trials are identical the conditions for a geometric distribution are ensured.

2 Use the geometric distribution function to find probabilities. The event $X > n$ is equivalent to zero sixes in n throws.

The intersection of the events $X > 30$ *and* $X > 20$ is $X > 30$

Exercise 1.3B Reasoning and problem-solving

Full A Level

1 A fair dice is thrown until a one appears for the first time and X is a random variable for the number of throws required.

a Write the probability distribution of X

b Find the probability that

i $X=4$ ii $X \geq 3$

c Calculate the mean and variance of X

2 A set of twelve cards is numbered 1–12. Cards are chosen at random with replacement until a number greater than 10 is obtained.

a Find the expected value of X, a random variable for the total number of cards chosen.

b What is the probability that a number greater than 10 occurs for the first time on the fourth pick?

3 An ordinary dice is thrown until a prime number appears. Find the mean and variance of X, the number of throws required.

4 In a sales promotion, 12% of packets of crisps contain a code which can be used to claim a prize. On each of five consecutive days you buy one packet of crisps.

a What is the probability that, of the five packets bought, exactly one prize is won?

b What is the probability that you win a prize for the first time on the 5th packet bought?

In each part you should define a random variable and give its distribution.

5 You stand in front of a shopping precinct and ask shoppers about their shopping habits. You find that, on average, you have to approach eight people until you meet one (the eighth person) who is prepared to answer your questions.

a Find the probability that any given person in the shopping precinct is prepared to answer questions. You should include details of any distribution used.

b Find also the probability that you have to approach at least three people in order to find someone who will answer questions.

6 Y is a random variable for the number of Bernoulli trials up to, *but not including*, the first success.

a Write the probability distribution of Y, specifying clearly its sample space.

b Find its expected value.

c Check your answer to the expected value of Y by using the relationship $Y=X-1$ where X is the number of trials *up to and including* the first success.

7 G is a random variable for the number of complete minutes that have passed until the first goal in a football match. Let p be the probability that any given one-minute interval includes a goal.

a Give two conditions which should be met so that G could be modelled by a geometric distribution.

b Assuming the conditions in part **a** to be met, write the probability distribution function of the random variable G.

c If $p=0.03$, find the probability that the first goal occurs between three and four minutes from the start of the match.

8 The random variable X has a geometric distribution, parameter p

a Prove that

$$P(X>n+m|X>m)=P(X>n)$$

b Explain how the result in part **a** shows that the geometric distribution is 'memoryless.'

1.4 The negative binomial distribution

Full A Level section

Fluency and skills

Full A Level

Given a sequence of independent trials, each with the same probability, p, of success, let the discrete random variable, X, represent the number of trials required to achieve the rth success.

X follows a negative binomial distribution with parameters r and p, the probability of success in a given trial. The geometric distribution is a special case of the negative binomial distribution with $r=1$

The negative binomial distribution can be used to model discrete random variables such as the number of throws of a dice up to the fourth 6 or the number of rocket launches up to the third failure.

Key point

If X is a negative binomial random variable with parameters r and p, then its probability distribution function is given by

$$\mathrm{P}(X=x)=\binom{x-1}{r-1}p^r(1-p)^{x-r};\ x=r,\ (r+1),\ \ldots$$

Recall that

$$\binom{n}{r}=\frac{n!}{r!(n-r)!}$$

The distribution is positively skewed and is defined for positive integer values of $x \geq r$. This graph shows the distribution for $r=3$, $p=0.6$

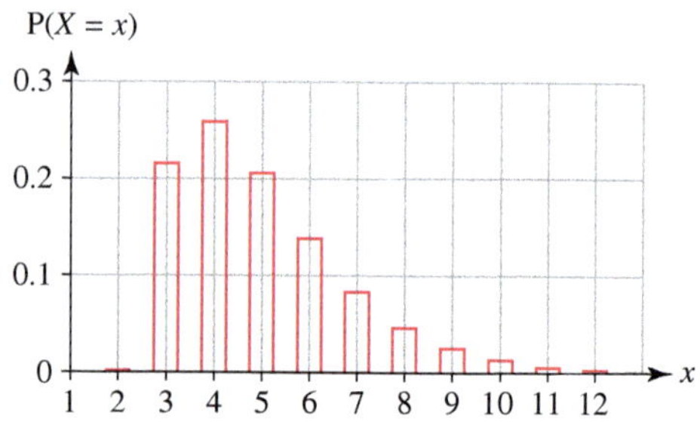

Example 1

Two independent observations are taken from a population with a negative binomial distribution, $r=3$, $p=0.6$. Find the probability that

a they are both greater than 5

b one is greater than 5 and the other is less than 4

a Let X be a random variable for any observation from the population.

$\mathrm{P}(X>5)=1-\mathrm{P}(X\leq 4)$

$=1-\{\mathrm{P}(X=3)+\mathrm{P}(X=4)\}$

$=1-(0.216+0.2592)=0.5248$

$\mathrm{P}(X\leq 4)=\mathrm{P}(X=3)+\mathrm{P}(X=4)$ because the possible values of X are 3 and 4.

P(both greater than 5) $=0.5248^2$

$=0.275$ (3sf)

The observations are independent.

b P(one is greater than 5 and the other is less than 4)

$=2\times 0.5248\times 0.216=0.227$ (3sf)

As you will see in the next section, the negative binomial distribution occurs in situations when you have the sum of r independent and identical geometric distributions. Since the mean and variance of a geometric distribution are $\frac{1}{p}$ and $\frac{1-p}{p^2}$ respectively, the following results are obtained.

Key point

A negative binomial random variable, X, has a mean and variance given by $\mu = E(X) = \frac{r}{p}$ and $\sigma^2 = Var(X) = \frac{r(1-p)}{p^2}$

Example 2

A random variable X has a negative binomial distribution with parameters $r = 4$ and $p = 0.5$

a Find the mean and variance of X

b An independent random sample of size 200 is taken from the population. How many observations would you expect to be within one standard deviation of the mean? Give your answer to the nearest integer.

a $E(X) = \frac{r}{p} = \frac{4}{0.5} = 8$ and $Var(X) = \frac{r(1-p)}{p^2} = \frac{4 \times 0.5}{0.5^2} = 8$

b Standard deviation of $X = \sqrt{8} = 2.828...$

P(within one standard deviation from the mean) $= P(6 \le X \le 10) = 0.6406$

Number out of 200 observations within one standard deviation from the mean is $200 \times 0.6406 = 128.12 = 128$ (to nearest integer)

Exercise 1.4A Fluency and skills

Full A Level

1 A random variable X has a negative binomial distribution, $r = 5$, $p = 0.7$. Find the probability that X takes a value

a equal to 9 **b** between 5 and 8 inclusive.

2 A random variable X has a negative binomial distribution with parameters $r = 3$, $p = \frac{4}{5}$

a Using the probability density function of X, copy and complete the table.

x	3	4	5	6	7	8
$P(X=x)$						

b Sketch the probability distribution function of X for $x = 3, 4, ..., 8$

3 **a** Write the probability distribution function for a random variable X with a negative binomial distribution, parameters $r = 5$, $p = 0.8$

b Write the mean and variance of X

c Find the values of

i $P(X = 6)$ **ii** $P(X \ge 3)$ **iii** $P(4 < X < 7)$

4 A random variable X has a negative binomial distribution, $r=6$, $p=0.6$

a Write the distribution function and the mean and variance of X

b Using the definition of the median of a discrete random variable as the smallest x such that $P(X \le x) \ge 0.5$, find the median of X

5 The random variable X has a negative binomial distribution with parameters r, p

a Write the probability distribution function of X

b Write the mean and variance of X

c For $r=4$ and $p=0.75$, use your answer in parts **a** and **b** to find the probability that X takes a value less than the mean.

6 a Write the probability distribution function and the mean and variance for a random variable X with a negative binomial distribution, parameters r and p

b Without calculation, explain why the distribution function for a negative binomial random variable should reduce to that of a geometric random variable when $r=1$

c Show that the results written in part **a** reduce to their geometric distribution equivalents when $r=1$

7 By using the negative binomial probability distribution function, show that

$P(X=r)=P(X=r+1)$ only if $r=\dfrac{1}{1-p}$

Reasoning and problem-solving

Full A Level

Strategy

To solve a problem using the negative binomial distribution

1. Ensure that the conditions for the distribution apply to the problem.
2. Use a calculator, statistical tables or the negative binomial distribution function to find probabilities.
3. Where required, use standard results to find the mean and variance of the distribution.

Suppose that a set of independent and identical Bernoulli trials was performed and each of the trials resulted in either 'success' or 'failure'. Let p be the probability of success in each trial and let X be a random variable for the number of trials up to and including the rth success.

The rth success on the xth trial means that the preceding $x-1$ trials contained $r-1$ successes and the rest were failures.

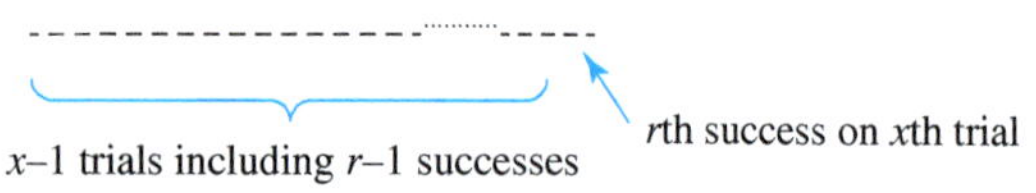

The series starts at $x=r$ because this is the minimum number of trials to achieve r successes.

Treating x as a constant, the probability of $x-1$ trials including $r-1$ successes is given by the binomial distribution function; that is,

$$\binom{x-1}{r-1}p^{r-1}(1-p)^{x-r};\, x=r,\,(r+1),\,\ldots$$

Hence, the probability that there are x trials up to and including the rth success is given by

$$P(X=x)=\binom{x-1}{r-1}p^{r-1}(1-p)^{x-r}p$$
$$=\binom{x-1}{r-1}p^{r}(1-p)^{x-r};x=r,\ (r+1),\ \ldots$$

This distribution can be thought of as the distribution for the waiting time to the rth success.

This is the distribution function of a negative binomial distribution mean $\frac{r}{p}$ and variance $\frac{r(1-p)}{p^2}$

Key point

The number of Bernoulli trials up to and including the rth success follows a negative binomial probability distribution, mean $\frac{r}{p}$ and variance $\frac{r(1-p)}{p^2}$

Notice that the distribution function and the mean and variance reduce to their geometric forms when $r=1$, as expected. See Exercise 1.4A, Question 6.

Example 3

In a fairground trial of strength, contestants use a hammer to attempt to hit a peg hard enough to trigger an alarm. It costs £2 for an unlimited number of attempts and three successes result in a prize of £10. A certain contestant has a probability of success in any given attempt of 0.3

a Write the distribution function of X, the number of trials up to and including the 3rd success.

b Find the probability of winning a £10 prize in at most five attempts.

c Calculate the mean number of attempts the contestant would need to achieve three successes.

d Explain why a negative binomial model may not be a good one.

a Let X be the number of trials to the 3rd success.

1 Ensure that the conditions for the distribution apply to the problem.

$X \sim$ Negative Binomial($r=3$, $p=0.3$) therefore,

$$P(X=x)=\binom{x-1}{2}0.3^3 0.7^{x-3};x=3,\ 4,\ \ldots$$

Use $P(X=x)=\binom{x-1}{r-1}p^r(1-p)^{x-r};$ $x=r,\ (r+1),\ \ldots$

b $P(X=3, 4 \text{ or } 5)=P(X=3)+P(X=4)+P(X=5)$
$=0.027+0.0567+0.07938$
$=0.16308$

2 Use the negative binomial distribution function to find probabilities.

c $E(X)=\frac{3}{0.3}=10$

3 Use standard results to find the mean of the distribution.

d The model assumes independence of outcomes from one attempt to the next and that trials are identical with no increase in skill, no tiring of the contestant etc. These are poor assumptions.

Exercise 1.4B Reasoning and problem-solving

Full A Level

1 A fair six-sided dice is thrown until a six has appeared three times. Let X be a random variable for the number of throws required.

   a Write the distribution of X and give its probability distribution function.

   b Find the mean and variance of X

   c Find the probability that 18 throws will be required.

2 One fifth of all bars of a certain brand of chocolate bar offer a free ticket to a concert. The chocolate bars cost 60p each and the concert tickets cost £5 each. Six friends wish to go to the concert and buy bars of the chocolate until they have six tickets. Let X be a random variable for the total number of bars they buy.

   a Write the distribution function of X.

   b Find the mean of X.

   c How much can the friends expect to pay using their strategy? Compare this with buying the tickets outright.

3 Satellites are placed in orbit using a rocket. In a large number of launches 8% of satellites are lost. Let X be the launch number of the 5th failure.

   a Write the probability distribution of X

   b Find the mean and variance of X

   c Find the probability of at least three failures in 50 launches. State any assumptions you make.

4 A certain type of childhood ear disorder can only be definitely diagnosed at surgery. 56% of children having an operation to diagnose the disorder actually have it.

   a What is the probability that the disorder is diagnosed

      i for the first time on the 4th operation,

      ii for the third time on the 6th operation?

   As part of a research project, it is necessary to find five children who definitely have the disorder.

   b What is the distribution function and the mean and variance of the number of operations required?

5 Suppose X and Y are independent random variables, each with a negative binomial distribution, parameters r, p and s, p respectively.

   a Interpret the random variable $X + Y$ in terms of trials and successes.

   b Write the distribution of $X + Y$

   c Find the mean and variance of $X + Y$

6 A fair coin is tossed repeatedly. Show that the probability that the 5th head will occur on the 8th toss given that the 12th head occurred on the 24th toss is $\dfrac{\binom{7}{4}\binom{15}{6}}{\binom{23}{11}}$

1.5 The Central Limit Theorem

Full A Level section

Fluency and skills

Full A Level

The Normal distribution can be used even if the underlying population from which a sample is taken is not Normally distributed. This is because of the celebrated **Central Limit Theorem** which says that, when sampling from a non-Normal distribution, the distribution of the sample mean tends to the Normal distribution as the sample size increases.

> When sampling from an underlying Normal population, the distribution of $\bar{X}$ is **exactly** Normal.

The diagrams below illustrate this idea for a discrete uniform distribution. A fair dice is thrown once. The probability distribution of X, the score on a randomly chosen throw, is shown. If many samples of size 2 are taken and the means of each sample calculated and plotted on a frequency diagram, the distribution acquires symmetry about a clearly defined peak (see $n = 2$ diagram). As the sample size increases further ($n = 5$, $n = 25$) the distribution of sample means tends to the characteristic bell-shaped curve with an increasingly reduced variance.

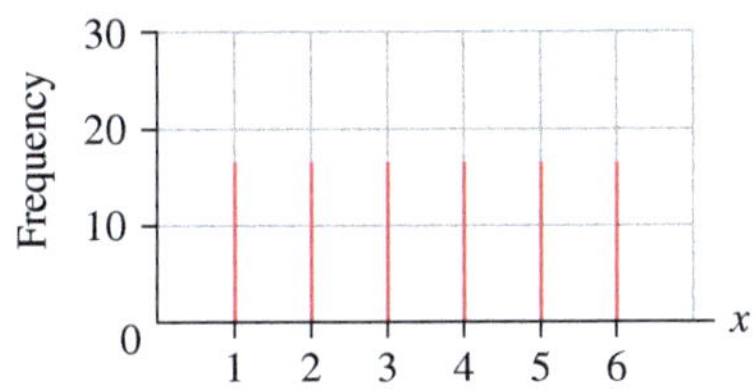

Discrete uniform distribution

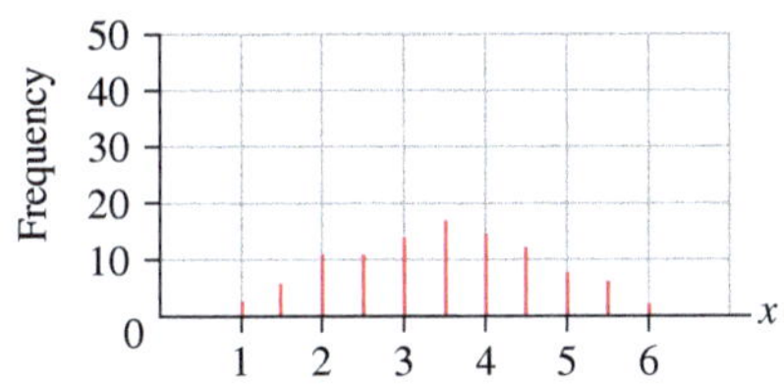

$n = 2$

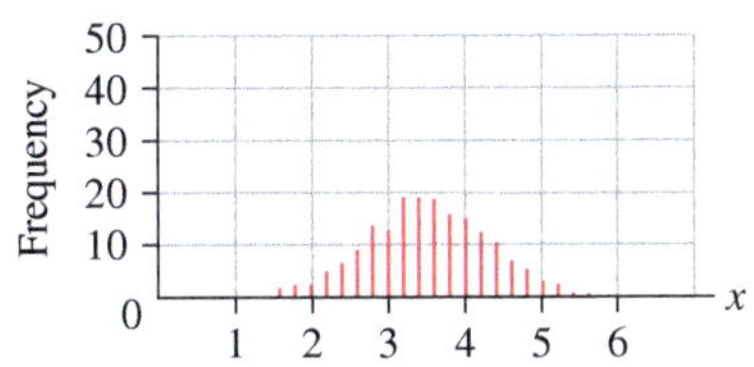

$n = 5$

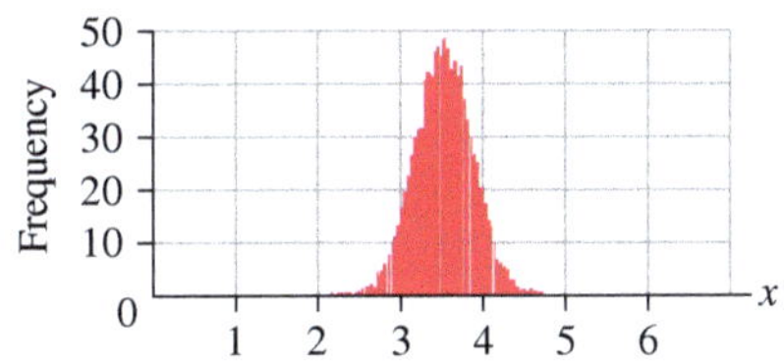

$n = 25$

Key point

Let X_i, $i = 1, 2, \ldots, n$ be an independent random sample from a population with mean μ and variance σ^2 and define the random variable $\bar{X} = \frac{1}{n}\sum_{i=1}^{n} X_i$

The distribution of $\bar{X}$ tends to $\mathrm{N}\left(\mu, \frac{\sigma^2}{n}\right)$ as n tends to infinity.

The Normal distribution provides a good approximation to the distribution of $\bar{X}$ for sample sizes of at least 30, fewer if the underlying distribution is fairly symmetric.

Example 1

$X_1, X_2, \ldots, X_{40}$ are independent and identically distributed discrete random variables each with mean 12 and standard deviation 3.

a Estimate the probability that $\bar{X}$, defined as $\frac{1}{40}\sum_{i=1}^{40} X_i$, takes a value between 11.5 and 12.5 inclusive.

b Explain why your answer in part **a** is an estimate.

a $\bar{X} \sim N(12, 0.225)$

Use $N\left(12, \frac{3^2}{40}\right)$

$P(11.5 \le \bar{X} \le 12.5) = 0.708$ (3sf)

Remember that for N(12, 0.225), the mean is 12 and the standard deviation is $\sqrt{0.225}$

b Distribution of $\bar{X}$ is only approximately Normal.

Since n is large, the Central Limit Theorem can be used.

Example 2

A fair 10-sided dice is repeatedly thrown and the random variable X denotes the score on a randomly chosen throw.

a Find the mean and variance of X

The random variable $\bar{X}_{30}$ denotes the mean of a randomly chosen sample of size 30.

b Give the approximate distribution of $\bar{X}_{30}$

c Find

i $P(\bar{X}_{30} \le 6)$ **ii** $P(\bar{X}_{30} > 4.5)$

a $$E(X) = 1 \times \frac{1}{10} + 2 \times \frac{1}{10} + \ldots + 10 \times \frac{1}{10}$$
$$= \frac{55}{10}$$
$$= 5.5$$

$$Var(X) = \left(1^2 \times \frac{1}{10} + 2^2 \times \frac{1}{10} + \ldots + 10^2 \times \frac{1}{10}\right) - 5.5^2$$
$$= \frac{33}{4}$$
$$= 8.25$$

b N(5.5, 0.275)

Use $N\left(5.5, \frac{8.25}{30}\right)$

c i 0.830 (3sf) **ii** 0.972 (3sf)

Exercise 1.5A Fluency and skills

Full A Level

1 A discrete random variable, X, has a uniform probability distribution with sample space consisting of integer values from 1 to 5 inclusive.

a Find the mean and variance of X and $P(X > 3.5)$

A sample of size 45 is taken and the sample mean calculated.

b Find the probability that this mean takes a value less than 2.8

2 A random variable, X, has a probability density function given by $P(X=x)=\begin{cases} \frac{x}{k}, & x=2,3 \\ 0, & \text{otherwise} \end{cases}$

a Prove that $k=5$ and find the mean and variance of X

A sample of size 35 is taken from the population and its mean, $\bar{X}$, calculated.

b Estimate the probability that $\bar{X}$ takes a value greater than 2.5

3 A random variable, X, has a probability density function given by $P(X=x)=\begin{cases} \frac{1}{a}x^2, & x=1,2,3,4 \\ 0, & \text{otherwise} \end{cases}$

a Prove that $a=30$

An independent random sample of 42 X-values is taken and the sample mean, $\bar{X}$, is calculated.

b Write the approximate distribution of $\bar{X}$ and find $P(\bar{X} < 3.2)$

4 A six-sided dice with faces numbered 1–6 is biased so that the probability that it lands on any particular value is proportional to that value.

a The dice is thrown once and X is a random variable for the score obtained. Find the mean and variance of X

b Find the probability that the mean score out of 32 throws is less than 4

5 A random variable X has distribution $P(X=x) = \frac{1}{4}, x = 1, 2, 3, 4$

a Find the mean and variance of X

A sample of size 2 is taken and the mean, $\bar{X}$, calculated.

b Find the (exact) probability distribution of $\bar{X}$ and $P(\bar{X} \geq 3)$

c Use a Normal approximation with a continuity correction to find $P(\bar{X} \geq 3)$

d Use your answers to parts **b** and **c** to comment on the use of the Central Limit Theorem for small samples.

Reasoning and problem-solving

Full A Level

Strategy

To solve a problem using the Central Limit Theorem

1. Check that the conditions for use of the Central Limit Theorem apply: non-Normal underlying population and large sample.
2. Calculate the mean and variance of the underlying population.
3. Write the approximate distribution of the sample mean, $\bar{X}$
4. Use Normal distribution theory/software to find probabilities as required.

The Central Limit Theorem can be applied to other distributions.

Example 3

A random sample of size 40 is taken from a binomial distribution with parameter $n = 8$, $p = 0.4$

a Find the (exact) probability that the first observation in the sample takes a value less than 3.5

b Estimate the probability that the sample mean takes a value less than 3.5

Let the sample be denoted by $X_1, X_2, \ldots, X_{40}$ and let $\bar{X} = \frac{1}{n}\sum_{i=1}^{n} X_i$

a $P(X_1 < 3.5) = 0.594$ (3 sf)

b Population mean and variance are given by $\mu = np = 3.2$ and $\sigma^2 = np(1-p) = 1.92$

Therefore $\bar{X}$ is approximately $N\left(3.2, \frac{1.92}{40}\right)$

$$P(\bar{X} < 3.5) = P\left(Z < \frac{3.5 - 3.2}{\sqrt{\frac{1.92}{40}}}\right) = P(Z < 1.37) = 0.915 \text{ (3sf)}$$

(2) Calculate the mean and variance of the underlying population.

(3) Write the approximate distribution of the sample mean $\bar{X}$

(4) Use Normal distribution theory to find probabilities as required.

There is no need to use a continuity correction here because of the large sample size.

Example 4

A Geiger counter detects radioactive disintegrations at a mean rate of 12 per minute.

a Write the distribution of X, the number of disintegrations in a randomly chosen four-minute interval, and its variance.

The number of disintegrations in each of 30 four-minute intervals is noted and the mean, $\bar{X}$, calculated.

b Write the approximate distribution of $\bar{X}$ and use it to find the probability that it will take a value greater than or equal to 46

a $X \sim Po(48)$; variance 48

b Sample variance $= \frac{\text{Variance of } X}{n} = \frac{48}{30} = 1.6$

$\bar{X} \sim N(48, 1.6)$ approximately

$P(\bar{X} \geq 46) = 0.943$ (3sf)

(2) Calculate the mean and variance of the underlying population.

(3) Write the approximate distribution of the sample mean $\bar{X}$

(4) Use Normal distribution theory/software to find probabilities as required.

Example 5

A random variable, X, has a probability distribution given by

$P(X = x) = (1-p)^{x-1}p; x = 1, 2, 3, \ldots$

a State the distribution of X and find its mean, μ, and variance, σ^2, if $p = 0.5$

A random sample of 30 X-values is taken and the mean, $\bar{X}$, of the sample calculated.

b Use the Central Limit Theorem to find the approximate probability that the mean is less than 2.1

a Geometric, parameter p. $\mu = \frac{1}{p} = 2$ and $\sigma^2 = \frac{(1-p)}{p^2} = 2$

b $\bar{X} \sim N\left(2, \frac{1}{15}\right)$ approximately

$P(\bar{X} < 2.1) = 0.651$ (3sf)

2 Calculate the mean and variance of the underlying population.

3 Write the approximate distribution of the sample mean $\bar{X}$

4 Use Normal distribution theory/software to find probabilities as required.

Example 6

In the manufacture of precision telescope mirrors, 15% are rejected at the final quality control stage. When five mirrors have been rejected, they are disposed of as a batch.

a Write the distribution function of X, the total number of mirrors manufactured up to and including the fifth rejection.

b Find the mean and variance of X

$\bar{X}$ is a random variable for the mean number of mirrors produced per batch up to the disposal of the 30th batch.

c Write the approximate distribution of $\bar{X}$ and calculate $P(\bar{X} \geq 25)$

a $X \sim$ Negative Binomial(5, 0.15)

b Mean $= \frac{r}{p} = \frac{5}{0.15} = 33.\dot{3}$ and

variance $= \frac{r(1-p)}{p^2} = \frac{5(1-0.15)}{0.15^2} = 188.\dot{8}$

c $\bar{X} \sim N\left(33.\dot{3}, \frac{188.\dot{8}}{30} = 6.296\right)$ approximately

$P(\bar{X} \geq 25) = 1 - 0.00045 = 0.99955$

2 Calculate the mean and variance of the underlying population.

3 Write the approximate distribution of the sample mean $\bar{X}$

4 Use Normal distribution theory/software to find probabilities as required.

Exercise 1.5B Reasoning and problem-solving

Full A Level

1 A random sample of size 30 is taken from a population with a binomial distribution, parameters $n=6$, $p=0.6$

a Write the approximate distribution of $\bar{X}$, the sample mean.

b Estimate the probability that the sample mean takes a value less than 4

2 The number of currants in a teacake follows a Poisson distribution with mean 12 per 50 millilitres.

a Write the distribution of X, the number of currants in a randomly chosen teacake of volume 250 ml.

42 teacakes are chosen at random, the number of currants in each is noted and the mean, $\bar{X}$, calculated.

b Write the approximate distribution of $\bar{X}$

c Find the probability that $\bar{X}$ will take a value less than or equal to 62

3 Let X_i; $i=1, 2, \ldots, n$ be an independent random sample of size n from a Bernoulli population; that is

$$P(X_i = a) = \begin{cases} p; a=1 \\ 1-p; a=0 \end{cases}; i=1, 2, \ldots, n$$

a Calculate the mean and variance of each of the X_i

The sample mean is given by $\bar{X} = \frac{1}{n}\sum_{i=1}^{n} X_i$

b Use the Central Limit Theorem to write the approximate distribution of $\bar{X}$ for large n

Let the random variable X_T be the sample total.

c Write the exact distribution of X_T

d Use your answer to part **b** and the result $X_T = n\bar{X}$ to write the approximate distribution of X_T. (This is the well-known Normal approximation to the binomial distribution.)

4 Cards are chosen at random with replacement from a 52-card deck until a queen appears. The random variable X is the number of cards chosen up to and including the queen.

a State the distribution of X and give its density function.

b Find the mean and variance of X

A random sample of 30 x-values is taken and the mean of the sample calculated.

c Use the Central Limit Theorem to find the probability that the mean is less than 13.5

5 A Geiger counter detects radioactive disintegrations at a mean rate of three per minute.

a Write the distribution of X, the number of disintegrations in 5 seconds.

b Find the probability that a randomly chosen 5-second period will contain

i one disintegration

ii more than one disintegration.

An experiment consists of counting the number of 5-second intervals up to and including the third disintegration.

c Write the approximate distribution and mean and variance of Y, the number of intervals required, stating why your answer to part **b ii** is relevant here.

The experiment is performed 30 times and the mean number of intervals until the third disintegration is calculated.

d Write the approximate distribution for this variable and calculate the probability that it takes a value less than 16.5

1.6 Probability generating functions

Full A Level section

Fluency and skills

Full A Level

A **probability generating function** can be defined for discrete random variables taking only integer values. It can be used to generate probabilities of the various values of the random variable and therefore provides a neat way of summarising the whole probability distribution. It can also be used to calculate the mean and variance of the corresponding probability distribution.

Consider a random variable X with probability distribution $P(X=x_i)=p_i;\ i=1, 2, \ldots, n$

Key point

The probability generating function, $G_X(t)$, of a random variable X is defined as $G_X(t)=E[t^X]$

The expectation of a function of a random variable is given by

$E[g(X)]=\sum_{\forall x} g(x)P(X=x)$

Therefore

$G_X(t)=E[t^X]=\sum_{i=1}^{n} p_i t^{x_i}$

For example, the coefficient of t^2 in a generating function is $P(X=2)$

Note that $G_X(1)=\sum_{i=1}^{n} p_i=1$, total probability equals 1

Example 1

A random variable Y has a probability distribution given by $P(Y=y)=\begin{cases} p; & y=1 \\ 1-p; & y=0 \end{cases}$

Find its probability generating function.

$G_X(t)=E[t^X]=\sum_{i=1}^{n} p_i t^{x_i}$ — Use the general definition of a probability generating function (pgf).

$G_Y(t)=t^0(1-p)+t^1 p$ — $G_Y(t)$ is the pgf of Y and so, by the definition of a pgf, this is equivalent to the statements $P(Y=1)=p$ and $P(Y=0)=1-p$

$=1-p+pt$

Example 2

A random variable X has a probability distribution $P(X=x) = kx^2;\ x=1, 2, 3$

Find the value of k and the generating function of X

$k+4k+9k=1$

so $k=\frac{1}{14}$

$G_X(t)=\frac{1}{14}t+\frac{2}{7}t^2+\frac{9}{14}t^3$

Total probability in sample space equals 1

The coefficient of t^x equals $P(X=x)$

Probability generating functions can be used to find the mean and variance of discrete distributions.

This result is proved in Exercise 1.6A, Question 7

Key point

If $G_X(t)$ is the probability generating function of a random variable X, and μ and σ^2 are the mean and variance of X, respectively, then $\mu = G_X'(1)$ and $\sigma^2 = G_X''(1)+\mu-\mu^2$ where G' and G'' are the first and second derivatives respectively of the function G.

Example 3

A random variable X has probability distribution

$$P(X=x)=\begin{cases} 0.3; x=1 \\ 0.4; x=2 \\ 0.3; x=3 \end{cases}$$

Write its probability generating function, $G_X(t)$, and hence find the mean and variance of X

$G_X(t)=0.3t+0.4t^2+0.3t^3$

$G_X'(t)=0.3+0.8t+0.9t^2$ so $\mu=G_X'(1)=2$

$G_X''(t)=0.8+1.8t$ so $\sigma^2=G_X''(1)+\mu-\mu^2=2.6+2-4=0.6$

Setting $t=1$ gives the value of the mean.

Setting $t=1$ gives an expression that can be used to find the variance.

Probability generating functions can be found for the standard discrete distributions and used to find the mean and variance of those distributions.

A binomial random variable, X, parameters n and p, has a probability distribution

$$P(X=x)=\binom{n}{x}p^x q^{n-x};\ x=0, 1, 2, \dots, n$$

where $q=1-p$

This gives a probability generating function

$$G_X(t) = E[t^X] = \sum_{x=0}^{n} \binom{n}{x} p^x q^{n-x} t^x$$

$$= \sum_{x=0}^{n} \binom{n}{x} (pt)^x q^{n-x}$$

Expanding gives

$$G_X(t) = q^n + \binom{n}{1}(pt)q^{n-1} + \ldots + (pt)^n$$

which you should recognise as the binomial expansion of $(q+pt)^n$

Therefore,

$$G_X(t) = (q+pt)^n$$

Differentiating $G_X(t)$ twice gives

$$G_X'(t) = n(q+pt)^{n-1} p$$

$$G_X''(t) = n(n-1)(q+pt)^{n-2} p^2$$

Since $q + p = 1$

$$G_X'(1) = np$$

$$G_X''(1) = n(n-1)(q+p)^{n-2} p^2$$

$$= n(n-1)p^2$$

Therefore,

$$\mu_X = G_X'(1) = np$$

and

$$\sigma_X^2 = G_X''(1) + \mu - \mu^2$$

$$= n(n-1)p^2 + np - (np)^2$$

$$= np - np^2$$

$$= np(1-p)$$

Key point

A binomial random variable, X, parameters n and p, with probability distribution

$$P(X = x) = \binom{n}{x} p^x q^{n-x}$$

has a probability generating function

$$G_X(t) = (q+pt)^n$$

and mean and variance

$\mu_X = np$ and $\sigma_X^2 = np(1-p)$

You are expected to be able to prove this and the next three key points. The proofs for these three results are outlined in Exercise 1.6A Questions 6 and 9 and Example 7.

Key point

A Poisson random variable, X, parameter λ, with distribution function

$$P(X=x)=\frac{e^{-\lambda}\lambda^x}{x!}; x=0,1,\ldots$$

has a probability generating function

$$G_X(t)=e^{\lambda(t-1)}$$

and mean and variance

$$\mu_X=\lambda \text{ and } \sigma_X^2=\lambda$$

Key point

A geometric random variable, X, parameter p, with distribution function

$$P(X=x)=(1-p)^{x-1}p; x=1,2,\ldots$$

has a probability generating function

$$G_X(t)=tp\left(\frac{1}{1-(1-p)t}\right)$$

and mean and variance

$$\mu_X=\frac{1}{p} \text{ and } \sigma_X^2=\frac{1-p}{p^2}$$

Key point

A negative binomial random variable, X, parameter r and p, with distribution function

$$P(X=x)=\binom{x-1}{r-1}p^r(1-p)^{x-r}; x=r,(r+1),(r+2),\ldots$$

has a probability generating function

$$G_X(t)=\left(\frac{tp}{1-(1-p)t}\right)^r$$

and mean and variance

$$\mu_X=\frac{r}{p} \text{ and } \sigma_X^2=\frac{r(1-p)}{p^2}$$

Example 4

Write the probability generating functions for these distributions.

i Binomial $(n=6, p=0.3)$

ii Geometric $(p=0.1)$

iii Poisson $(\lambda=4)$

i $(0.7+0.3t)^6$

ii $0.1t\left(\frac{1}{1-0.9t}\right)$

iii $e^{4(t-1)}$

Example 5

A Bernoulli random variable, B, has a probability distribution given by

$$P(B=b)=\begin{cases} p; & b=1 \\ 1-p; & b=0 \end{cases}$$

Show that its probability generating function is $G_B(t)=1-p+pt$ and use this to show that the mean and variance of B are given by $\mu_B=p$ and $\sigma_B^2=p(1-p)$

$G_B(t)=E[t^B]$
$=t^1p+t^0(1-p)$
$=1-p+pt$

$G_B'(t)=p$ and $G_B''(t)=0$

$G_B'(1)=p$ and $G_B''(1)=0$

$\mu_B=G_B'(1)=p$ as required

$\sigma^2=G_B''(1)+\mu_B-\mu_B^2$
$=0+p-p^2$
$=p(1-p)$ as required

Exercise 1.6A Fluency and skills

Full A Level

1 A random variable X has probability distribution $P(X=x)=\begin{cases} 0.3; x=1 \\ 0.4; x=2 \\ 0.3; x=3 \end{cases}$

Write its probability generating function.

2 A pack of 14 cards contains five 1s, five 2s and four 4s. A card is chosen at random and X denotes the value of that card. What is the probability generating function for X?

3 A random variable S has probability distribution $P(S=s)=ks^2; s=1, 2, 3, 4$

a Find the value of k

b Use your value of k to find the probability generating function of S. Hence calculate the mean and variance of S

4 A random variable R has probability distribution $P(R=r)=\frac{1}{k}r; r=2, 3, 4, 5$

a Write the generating function of R in terms of k

b Use your answer to find the value of k and hence calculate the mean and variance of R

5 A geometric random variable X with parameter p has a probability generating function given by

$G_X(t)=\frac{t}{4-3t}$

a Find p

b Write the distribution function of X and use it to find the probabilities that $X=1$, 2 and 3

c By differentiation of $G_X(t)$, find the mean and variance of X

6 A Poisson random variable, X, has parameter λ. Prove that its probability generating function is $G_X(t) = e^{\lambda(t-1)}$ and its mean and variance are given by $\mu_X = \lambda$ and $\sigma_X^2 = \lambda$

7 A random variable X has distribution function $P(X = x_i) = p_i; i = 1, 2, ..., n$

a Show that $G_X(t) = \sum_{i=1}^{n} p_i t^{x_i}$

b Prove that $G_X(1) = 1$

c Prove that $G'_X(1) = \sum_{i=1}^{n} p_i x_i$ and $G''_X(1) = \sum_{i=1}^{n} p_i x_i (x_i - 1)$

d Hence write an expression for μ in terms of its probability generating function and prove directly that $\sigma^2 = G''_X(1) + \mu - \mu^2$

8 A bag contains one red and four green balls. Balls are taken with replacement, one at a time, until a red ball is drawn. Y is a random variable for the number of draws *up to but not including* the final draw. Find the probability generating function of Y and use it to find $E[Y]$

9 A geometric random variable, X, has distribution function $P(X = x) = (1-p)^{x-1} p$. Prove that its probability generating function is $G_X(t) = tp\left(\frac{1}{1-qt}\right)$, $(q = 1 - p)$ and its mean and variance are given by $\mu_X = \frac{1}{p}$ and $\sigma_X^2 = \frac{q}{p^2}$

10 A discrete uniform random variable, X, has a probability distribution $P(X = x) = \frac{1}{a}; x = 1, 2, ..., a$

By finding the probability generating function of X, prove that the mean of X is given by $\mu = \frac{a+1}{2}$

11 A random variable, S, has a probability generating function given by $G_S(t) = \frac{1}{a}(1 + t + 3t^2 + bt^3)$ where a and b are constants.

a Given that $E[S] = \frac{5}{3}$, show that $a = 6$ and $b = 1$

b Using your answer in part **a**, find the variance of S

Reasoning and problem-solving

Full A Level

To solve a problem using probability generating functions

1. Use $G_X(t) = E[t^X]$ to find the probability generating functions of random variables.
2. Use differentiation to find expressions for the mean and variance in terms of $G_X(t)$
3. Calculate the mean and variance for standard distributions.
4. Find the probability generating function for $X + Y$ where these are independent random variables with known generating functions.

The probability generating function for a sum of independent random variables can be found from the individual generating functions.

Key point

Consider two independent random variables, X and Y, with generating functions $G_X(t)$ and $G_Y(t)$, respectively. The generating function of $X + Y$ is given by $G_{X+Y}(t) = G_X(t) \cdot G_Y(t)$

This is proved in Exercise 1.6B Question 8

Example 6

X and Y are two independent binomial random variables with distributions $X \sim B(10, 0.25)$ and $Y \sim B(10, 0.5)$

a Find the probability generating function (pgf) of the variable $X + Y$

b Use the pgf to prove that the mean of $X + Y$ is 7.5

a $G_X(t) = (0.75 + 0.25t)^{10}$ and $G_Y(t) = (0.5 + 0.5t)^{10}$

b $G_{X+Y}(t) = G_X(t).G_Y(t)$

$= (0.75 + 0.25t)^{10} \times (0.5 + 0.5t)^{10}$

$= (0.125t^2 + 0.5t + 0.375)^{10}$

$G'_{X+Y}(t) = 10(0.125t^2 + 0.5t + 0.375)^9 \times (0.25t + 0.5)$

$G'_{X+Y}(1) = 10 \times 1 \times 0.75$

$= 7.5$ as required.

1 Find the probability generating functions of random variables. These are standard results for a binomial distribution.

4 Find the probability generating function for $X + Y$.

2 Use differentiation to find an expression for the mean in terms of $G_X(t)$.

A further result follows if the random variables are independent and identically distributed (i.i.d.).

Key point

If X_1 and X_2 are independent and identically distributed random variables each with generating function $G_X(t)$ then the generating function of $X_1 + X_2$ is given by $G_{X_1+X_2}(t) = (G_X(t))^2$

This result can be generalised to apply to an independent random sample of size n (see Exercise 1.6B Question 7).

Example 7

By using the probability generating function of a geometric random variable, parameter p, show that the probability generating function of a negative binomial random variable, parameters r and p, is given by $G_X(t) = \left(\frac{tp}{1-qt}\right)^r$

Let $X \sim$ Geometric(p) and X_i; $i = 1, 2, \ldots, r$ be an independent random sample from the geometric distribution. $\sum_{i=1}^{r} X_i \sim$ Negative Binomial(r, p)

But $\sum_{i=1}^{r} X_i$ has a probability generating function given by $\left(\frac{tp}{1-qt}\right)^r$

Therefore the probability generating function of a negative binomial random variable, parameters r and p, is given by $\left(\frac{tp}{1-qt}\right)^r$

Sums of i.i.d. geometric random variables are negative binomial.

Using the general result.

Example 8

You have two packs of cards, the first a pack of eight cards numbered 1–8 and the second with six cards numbered 1–6. You sample with replacement from each of the packs until you get a six. Find the pgf and hence the mean of T, the total number of cards drawn including the two sixes.

Let R and S be random variables for the number of draws required from each of the two packs.

$R \sim \text{Geometric}\left(\frac{1}{8}\right)$ and $S \sim \text{Geometric}\left(\frac{1}{6}\right)$

$$G_T(t) = t^2 \frac{1}{8} \times \frac{1}{6}\left(\frac{1}{1-\frac{7}{8}t}\right)\left(\frac{1}{1-\frac{5}{6}t}\right)$$

$$= t^2 \times \frac{1}{8-7t} \times \frac{1}{(6-5t)}$$

$$= \frac{t^2}{48-82t+35t^2}$$

Find the probability generating functions of random variables. This is the pgf of T because $T = R + S$.

$$G_T'(t) = \frac{96t-82t^2}{(48-82t+35t^2)^2}$$

Use differentiation to find expressions for the mean in terms of G(t).

$G_T'(1) = 14$

Therefore $E[T] = 14$

Exercise 1.6B Reasoning and problem-solving

Full A Level

1 X and Y are two independent Poisson-distributed random variables with means 3 and 4, respectively.

a Given that the probability generating function of a Poisson random variable parameter μ is given by $G(t) = e^{-\mu(1-t)}$, write and simplify the probability generating function for the random variable $X + Y$

b Hence state the distribution of $X + Y$

c State the general principle suggested by the result in part **b**.

2 A fair six-sided dice and a fair eight-sided dice are thrown. Prove that the probability generating function of X_T, the total score, is given by $G_X(t) = \dfrac{t^2(1-t^6)(1-t^8)}{48(1-t)^2}$

3 The dice in Question **2** are each thrown until a 1 shows.

a Prove that the pgf for the total number of throws required, T, is given by

$$G_T(t)=t^2\left\{\left(\frac{1}{6-5t}\right)\left(\frac{1}{8-7t}\right)\right\}$$

b Show that $G_T(1)=1$ and explain why this should be the case.

c Calculate the mean number of throws required.

4 Two independent observations are taken from a geometric distribution, $p=0.6$

a Find the probability generating function for the sum of these observations.

A single observation is taken from a negative binomial distribution, parameters $r=2$, $p=0.6$

b Find the probability generating function for this observation.

c Explain the relationship between your answers to parts **a** and **b**.

5 A fair dice is thrown three times. Treating this as an independent random sample of size three from a uniform distribution on the integers 1–6, show that the generating function of the sum of the scores is $\frac{1}{216}\times\frac{(t(1-t^6))^3}{(1-t)^3}$

6 X_1 and X_2 are two independent random variables, each following a Bernoulli distribution with parameter 0.4

a Find the probability generating function for the variable X_T, defined by $X_T = X_1 + X_2$

b Hence name the distribution of X_T and by differentiation, find its mean and variance.

7 X_i; $i=1, 2, \ldots, n$ are i.i.d. random variables each following a Bernoulli distribution, parameter p

a If $X_T = \sum_{i=1}^{n} X_i$, write the probability generating function of X_T

b Specify fully the distribution of X_T and use your answer to part **a** to find its mean and variance. (This is a general version of Question **6** and is a proof of the mean and variance of a binomial distribution.)

8 Two independent random variables, X and Y, have generating functions $G_X(t)$ and $G_Y(t)$ respectively. Prove that the generating function of $X+Y$ is given by $G_{X+Y}(t) = G_X(t)\times G_Y(t)$ (You should use that, for independent random variables, $E(XY) = E(X)\times E(Y)$)

1 Summary and review

Chapter summary

- The expected value or mean value and variance of a discrete random variable are given by

  $\mu = \mathrm{E}[X] = \sum_{\forall x} x\mathrm{P}(X = x)$

  $\sigma^2 = \mathrm{Var}(X) = \mathrm{E}[(X - \mu)^2] = \sum_{\forall x}(x - \mu)^2\mathrm{P}(X = x) = \sum_{\forall x} x^2\mathrm{P}(X = x) - \{\mathrm{E}(X)\}^2$

- The expected value and variance of a function, g(X), of a discrete random variable are given by

  $\mathrm{E}[\mathrm{g}(X)] = \sum_{\forall x}\mathrm{g}(x)\mathrm{P}(X = x)$

  $\mathrm{Var}(\mathrm{g}(X)) = \mathrm{E}(\{\mathrm{g}(X)\}^2) - \{\mathrm{E}(\mathrm{g}(X))\}^2 = \sum_{\forall x}\{\mathrm{g}(x)^2\}\mathrm{P}(X = x) - \{\mathrm{E}(\mathrm{g}(X))\}^2$

- If X and Y are two random variables and $Y = aX + b$, where a and b are constants, then
  - $\mathrm{E}(Y) = a\mathrm{E}(X) + b$ and $\mathrm{Var}(Y) = a^2\mathrm{Var}(X)$
  - the expected value of the sum, $X + Y$, is given by $\mathrm{E}(X + Y) = \mathrm{E}(X) + \mathrm{E}(Y)$
  - the variance of the sum, $X + Y$, is given by $\mathrm{Var}(X + Y) = \mathrm{Var}(X) + \mathrm{Var}(Y)$

- If X is the number of random events that occur in a given interval of time or space and λ is the mean number of occurrences in that interval then its probability distribution function is given by

  $\mathrm{P}(X = x) = \dfrac{\mathrm{e}^{-\lambda}\lambda^x}{x!};\ x = 0, 1, 2\ldots$

  and X is said to follow a Poisson distribution, parameter λ

- If X has a probability distribution function given by $\mathrm{P}(X = x) = \dfrac{\mathrm{e}^{-\lambda}\lambda^x}{x!};\ x = 0, 1, 2\ldots$

  then the mean and variance of X are respectively given by $\mathrm{E}(X) = \lambda$ and $\mathrm{Var}(X) = \lambda$

- If X_i; $i = 1, 2, \ldots, n$ is a set of n independent and identically distributed random variables each with a Poisson distribution, mean λ and if $X_T = \sum_{i=1}^{n} X_i$, then X_T has a Poisson distribution with mean $n\lambda$

- If events occur randomly at a rate of λ per given interval of time or space, then the number of events in n such intervals is a Poisson random variable with mean $n\lambda$

- If $X \sim \mathrm{B}(n, p)$ then X has an approximate Poisson distribution, parameter np, for large n and small p

Full A Level

- If X is a geometric random variable with parameter p then its probability distribution function is given by $P(X=x)=(1-p)^{x-1}p;\ x=1, 2, \dots$
- A geometric random variable, X, with distribution function $P(X=x)=(1-p)^{x-1}p;\ x=1, 2, \dots$ has a mean and variance given by $\mu=E[X]=\frac{1}{p}$ and $\sigma^2=\text{Var}[X]=\frac{1-p}{p^2}$
- The number of Bernoulli trials up to and including the first success follows a geometric probability distribution, mean $\frac{1}{p}$ and variance $\frac{1-p}{p^2}$
- If X is a negative binomial random variable with parameters r and p, then its probability distribution function is given by $P(X=x)=\binom{x-1}{r-1}p^r(1-p)^{x-r};\ x=r,\ (r+1), \dots$
- A negative binomial random variable, X, has a mean and variance given by $\mu=E(X)=\frac{r}{p}$ and $\sigma^2=\text{Var}(X)=\frac{r(1-p)}{p^2}$
- The number of Bernoulli trials up to and including the rth success follows a negative binomial probability distribution, mean $\frac{r}{p}$ and variance $\frac{r(1-p)}{p^2}$
- Let X_i, $i=1, 2, \dots, n$ be an independent random sample from a population with mean μ and variance σ^2 and define the random variable $\bar{X}=\frac{1}{n}\sum_{i=1}^{n}X_i$. The distribution of $\bar{X}$ tends to $N\left(\mu, \frac{\sigma^2}{n}\right)$ as n tends to ∞
- The probability generating function, $G_X(t)$, of a random variable X is defined as $G_X(t)=E[t^X]$
- If $G_X(t)$ is the probability generating function of a random variable X, and μ and σ^2 are the mean and variance of X, respectively, then $\mu=G_X'(1)$ and $\sigma^2=G_X''(1)+\mu-\mu^2$
- A binomial random variable, X, parameters n and p, with probability distribution $P(X=x)=\binom{n}{x}p^x q^{n-x};\ x=1, 2, \dots, n$ has a probability generating function $G_X(t)=(q+pt)^n$
- A Poisson random variable, X, parameter λ, with distribution function $P(X=x)=\frac{e^{-\lambda}\lambda^x}{x!}$ has a probability generating function $G_X(t)=e^{\lambda(t-1)}$
- A geometric random variable, X, parameter p, with distribution function $P(X=x)=(1-p)^{x-1}p$ has a probability generating function $G_X(t)=tp\left(\frac{1}{1-(1-p)t}\right)$
- A negative binomial random variable, X, parameters r and p, with distribution function $P(X=x)=\binom{x-1}{r-1}p^r(1-p)^{x-r};\ x=r, (r+1), (r+2), \dots$ has a probability generating function $G_X(t)=\left(\frac{tp}{1-qt}\right)^r$

Check and review

You should now be able to...	Review Questions
✓ Define and calculate expectation values for given random variables.	1
✓ Apply Poisson probability models in different circumstances.	2, 3
✓ Apply geometric and negative binomial probability models in different circumstances. Full A Level	3, 4, 5
✓ Apply the Central Limit Theorem when sampling from non-Normal populations.	6
✓ Derive probability generating functions for given random variables and use them to calculate corresponding means and variances.	7

1 Integers 1–6 are generated so that the probability of any given integer being chosen is proportional to its cube. Find the expected value of S, the score on a randomly chosen throw.

2 A study of World War II flying bomb impacts in South London shows that the average number of hits per $0.25\,\text{km}^2$ was 0.9323, based upon $144\,\text{km}^2$.

  a Assuming a Poisson distribution, calculate the expected number of $0.25\,\text{km}^2$ areas with two hits.

  b Comment on the spatial distribution of hits over South London given that the actual number of $0.25\,\text{km}^2$ areas with two hits was 93

3 A Geiger counter detects radioactive disintegrations at a mean rate of three per minute.

  a Use the Poisson distribution function to find the probability that a randomly chosen five-second interval will contain at least one disintegration.

  b Find the (exact) probability that the third disintegration will occur in the 12th five-second interval.

Full A Level

  c Using the negative binomial distribution, calculate the approximate value of the mean and variance of the number of five-second intervals up to and including the 3rd disintegration.

  d Use your answer in part **a** to explain why your answers in part **c** are approximate.

4 The random variable X has a geometric distribution, parameter p. Prove that $P(X > a) = q^a$ where $q = 1 - p$

5 A random variable X has a geometric distribution with parameter p

  a Find the probability that the number of trials up to and including the first success strictly exceeds k

  b Given that the first m trials do not include a success, find the probability that the number of further trials to the first success exceeds k

6 In a fairground game, a large tub is filled with red and green sweets. 20% of the sweets are red. A player choses sweets at random until a red sweet is chosen, and the tub is re-filled so that the proportion of sweets that are red remains constant at 20%. These sweets are packaged and given to the player. This is repeated until the player has received four packs of sweets.

a Find the mean and variance of X, the total number of sweets in the four packs.

40 players are observed and the mean of the 40 x-values calculated.

b Write the approximate distribution for this mean value and find P(mean ≥ 17)

7 **a** Prove that the probability generating function for a binomial-distributed random variable, parameters n and p, is $(q+pt)^n$ where $q = 1 - p$

b Use your result in part **a** to show that $P(S) = 1$, where S is the sample space for a binomial random variable.

Full A Level

Exploration

Going beyond the exams

Did you know?

The Poisson distribution is named after the French Mathematician, Geometer and Physicist Simeon Poisson.

Poisson is also well known for his partial differential equation, frequently written as $\nabla^2 \phi = f$, and widely used in mechanical engineering and theoretical physics.

> **Life is good for only two things: doing mathematics and teaching it.**
>
> \- Simeon Poisson

ICT

Many discrete distributions can be simulated in spreadsheet packages. For instance, the distribution of 100 rolls of a fair dice can be simulated in Excel.

Use the function =RANDBETWEEN(1,6) to generate a random number between 1 and 6. Then copy this formula into 100 cells in the A column. Use the COUNTIF function to total how many of each number appears in the list. (An example formula can be seen in cell D3). Then draw a graph of the results.

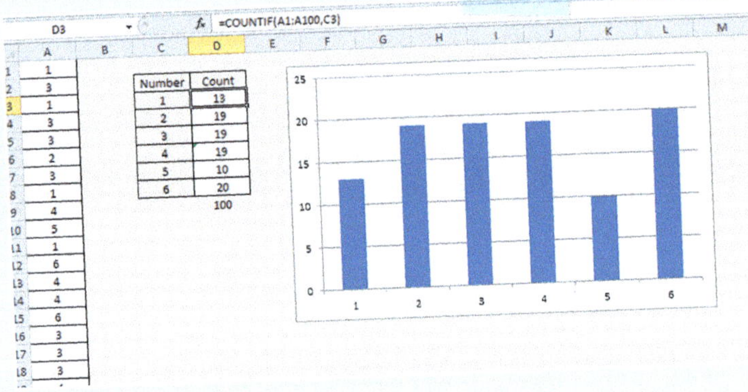

Now have a go at simulating the following situations using a similar process:

- 100 rolls on a twelve sided dice;
- 100 rolls on two six sided dice;
- 100 rolls on a biased 6 sided dice (for example, even numbers are twice as likely as odd numbers).

ICT

The Excel function =POISSON(x, mean, cumulative) can be used to calculate the probability of x events, with the mean average number of events per interval.

For example, by 2017 the mean average number of goals Ipswich Town scored against Norwich City per match was 1.455. Find the probability that

a. Ipswich Town score 5 goals in one game against Norwich City.

b. Ipswich Town score 4 or fewer goals in one game against Norwich City.

Using Excel,

a. =POISSON(5, 1.455, false) = 0.0127 (to 3 significant figures)

b. =POISSON(4, 1.455, true) = 0.983 (to 3 significant figures).

1 Assessment

1 The probability distributions of X and Y are as follows

$$P(X=x)=\begin{cases}\frac{x}{k};\ x=1,2,3,4\\ 0;\ \text{otherwise}\end{cases}$$

y	1	2	3	4
$P(Y=y)$	0.3	a	a	b

Given that $P(X=4)=P(Y=4)$

a Find the value of k, a and b **[5]**

b Which discrete random variable has the highest value on average?

Explain how you know. **[4]**

2 This table shows the probability distribution of N, the number of people living in a household as estimated from a sample of 200 households.

n	1	2	3	4	5
$P(N=n)$	$3k$	$8k$	$6k$	$2k$	k

a Find the value of k **[2]**

b Calculate the median and mode of N **[2]**

c Calculate the expected number of people per household. **[2]**

d Calculate the variance of N **[3]**

3 A shop sells apples at the rate of 2.5 per minute on average.

a Explain what assumptions you would have to make in order to model the sale of apples as a Poisson distribution. **[2]**

b Use the Poisson distribution to find the probability that in a one-minute period

i no apples are sold

ii fewer than three apples are sold

iii more than four apples are sold. **[6]**

c Calculate the probability that over a three-minute period, more than ten apples are sold. **[3]**

4 The discrete uniform variable U is defined for integers in the interval [2, 7]

a Calculate $E(U)$ and $Var(U)$ **[3]**

b Find $P(U>3.5)$ **[2]**

5 The discrete random variable V has pdf $f(v)=\begin{cases}\frac{v^2}{30}; x=1,2,3,4\\ 0; \text{otherwise}\end{cases}$

a Calculate the median of V **[2]**

b Calculate $E(5V+3)$ and $Var(5V+3)$ **[7]**

6 The random variable X, which takes integer values, is such that $X \sim Po(5.5)$

a Calculate the range of values of a such that $P(X<a)>0.7$ **[2]**

b Calculate the range of values of b such that $P(X>b)<0.1$ **[2]**

c Calculate the probability that the value of X is within one standard deviation of the mean. **[5]**

A second, independent, random variable Y is such that $Y \sim Po(2)$

d Calculate $P(X+Y<9)$ **[2]**

7 The discrete random variable T has pdf $f(t)=\begin{cases}\frac{1}{A}; t=1,2,3\\ \frac{t}{A}; t=4,5\\ 0; \text{otherwise}\end{cases}$

a Calculate $P(T>E(T))$ **[5]**

b Find $Var(7-3T)$ **[4]**

8 The probability density function of X is given by $f(x)=\begin{cases}\frac{1}{n}; x=1,2,\ldots,n\\ 0; \text{otherwise}\end{cases}$

a Write an expression for $E(X)$ **[1]**

b Prove that the variance of X is given by the formula $Var(X)=\frac{1}{12}(n^2-1)$ **[7]**

9 The number of emails a person receives between 9 am and 5 pm is modelled by a Poisson distribution. Emails are received at a rate of three every ten minutes. Every time period in this question falls between 9 am and 5 pm.

a Calculate the probability of receiving more than seven emails in a ten minute period. **[3]**

b Calculate the probability of receiving between five and eight emails (inclusive) over half an hour. **[3]**

c Calculate the probability of receiving at least three emails every ten minutes for an hour. **[5]**

10 A tetrahedral dice has sides numbered 1, 2, 3 and 4. The random variable X~Geometric(0.25) models the number of rolls required before the dice shows a 1

a Calculate

i $P(X \geq 3)$ **ii** $P(X \geq 5)$ **iii** $P(X \geq 5 | X \geq 2)$ **[5]**

b Explain why your answers to parts **a i** and **a iii** are equal. **[1]**

11 A gambling game involves throwing a basketball into a hoop until you miss. A thrower gets the ball into the hoop with probability 0.6. The random variable X describes the number of shots they take in total.

a State the distribution of X **[2]**

b State $E(X)$ **[1]**

c Find $P(X \geq E(X))$ **[3]**

12 A triple-elimination tournament is one where each competitor plays matches until they have lost three matches.

a State the distribution of X, the number of matches a competitor plays before they are eliminated. Assume that all players are equally skilled. **[2]**

b How many rounds should a competitor expect to play in the tournament? **[1]**

c Calculate $\text{Var}(X)$ **[2]**

d Calculate the probability that the number of rounds the competitor plays in the tournament is within one standard deviation of the expected number. **[3]**

13 For a discrete random variable X, the probability generating function is $G_X(z) = \sum_{x=0}^{\infty} P(X = x) \times z^x$

a Find the probability generating function for X~Po(λ) **[3]**

$\left(\text{Hint: Use the identity } e^z = \sum_{i=0}^{\infty} \frac{z^i}{i!}\right)$

b Use the probability generating function of X~Po(λ) to verify that $E(X) = \text{Var}(X) = \lambda$ **[4]**

14 a A random variable X is distributed X~Negative Binomial$\left(1, \frac{1}{4}\right)$ **[6]**

i State $P(X = k)$

ii Using the identity $\sum_{k=0}^{\infty} x^k = \frac{1}{1-x}$, show that the probability generating function is $\frac{z}{4-3z}$

iii Differentiate the pgf and hence prove that $E(X) = 4$

b A random variable Y, independent of X, is distributed Y~Negative Binomial$\left(2, \frac{1}{4}\right)$ **[7]**

i State $P(Y = k)$

Full A Level

ii Using the identity $\sum_{k=0}^{\infty} kx^k \equiv \frac{x}{(1-x)^2}$ show that the probability generating function is $\frac{z^2}{(4-3z)^2}$

iii Differentiate the pgf and hence prove that $E(Y) = 8$

c Explain why, if $Z \sim$ Negative Binomial$\left(3, \frac{1}{4}\right)$, $E(Z) = E(X) + E(Y)$. **[1]**

15 A negative binomial random variable counts the number of successes recorded in a series of trials before r failures occur, where the probability of a success is p. For a particular negative binomial random variable, X, the probability generating function is

$$G_X(z) = \sum_{k=0}^{\infty} (k+1)\left(\frac{1}{5}\right)^k \left(\frac{4}{5}\right)^2 z^k$$

a Determine r and p **[2]**

b Calculate $E(X)$ and $Var(X)$ **[3]**

c Find $P(|X - E(X)| < 3\,Var(X))$ **[5]**

16 The geometric distribution counts the number of trials before a single success occurs. For a random variable, X, which follows the geometric distribution the probability generating function can be written as $G_X(z) = \sum_{k=0}^{\infty} \frac{0.25}{0.75^{1-k}} z^k$

a Determine p, the probability of a success. **[1]**

b Calculate the mean and variance of X **[2]**

A different geometrically distributed random variable, Y, has parameter $p = 0.4$

c Calculate the mean and variance of Y and compare them with the corresponding statistics for X **[3]**

Somebody claims that they can find a geometrically distributed random variable, $Z \sim Geo(p)$, whose mean is larger than Y's but whose variance is smaller.

d Show that both $E(Z)$ and $Var(Z)$ are strictly decreasing functions of p and hence prove that the claim cannot be correct. **[4]**

17 Two independent identical Normal distributions are $X, Y \sim N(5, 16)$

a State the distributions of $X + Y$ and $X - Y$ **[4]**

b Calculate

i $P(X < 3)$ ii $P(Y < 2)$ iii $P(X + Y < 5)$ **[4]**

2 Hypothesis testing and association testing

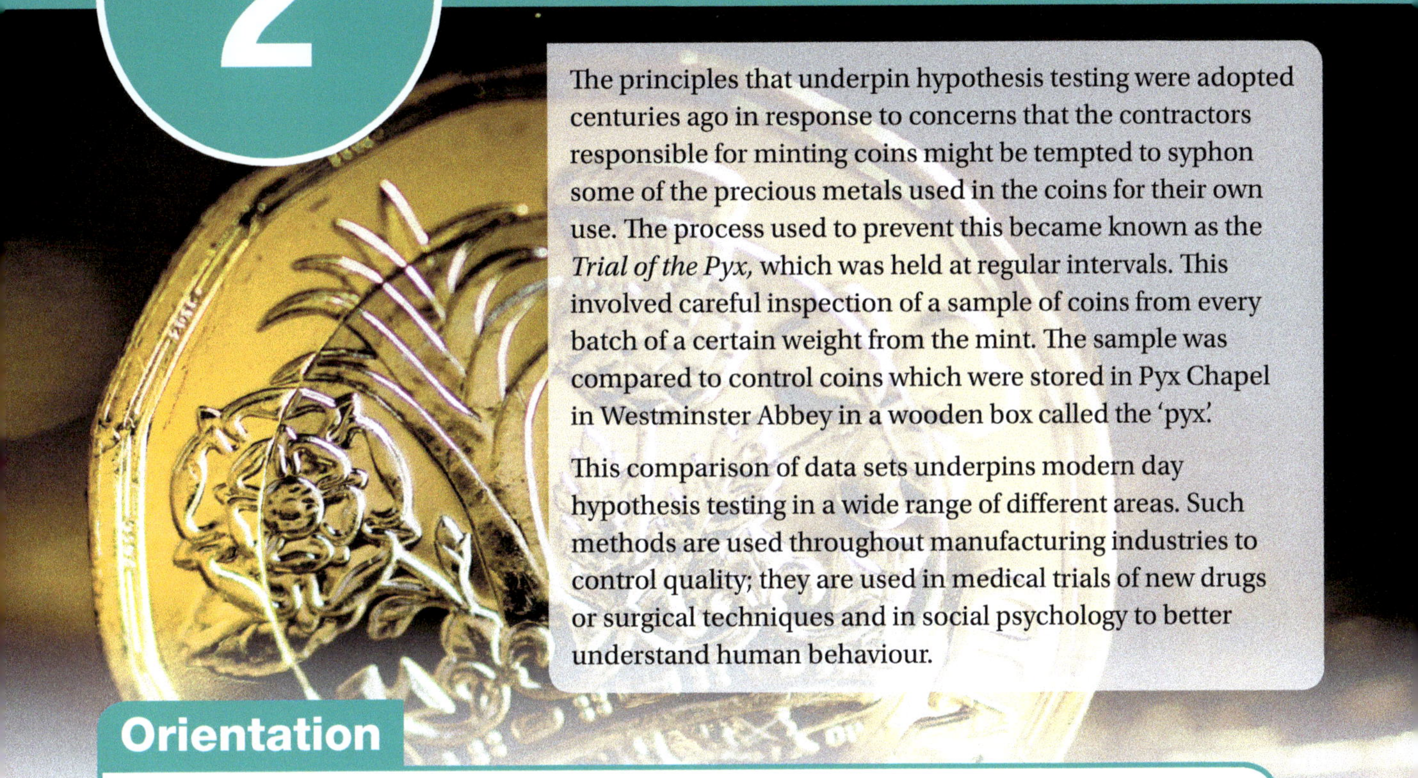

The principles that underpin hypothesis testing were adopted centuries ago in response to concerns that the contractors responsible for minting coins might be tempted to syphon some of the precious metals used in the coins for their own use. The process used to prevent this became known as the *Trial of the Pyx,* which was held at regular intervals. This involved careful inspection of a sample of coins from every batch of a certain weight from the mint. The sample was compared to control coins which were stored in Pyx Chapel in Westminster Abbey in a wooden box called the 'pyx'.

This comparison of data sets underpins modern day hypothesis testing in a wide range of different areas. Such methods are used throughout manufacturing industries to control quality; they are used in medical trials of new drugs or surgical techniques and in social psychology to better understand human behaviour.

Orientation

What you need to know

Maths Ch11

- Conduct and interpret a hypothesis test.

What you will learn

- To test the parameter of a Poisson distribution.
- To test the parameter of a geometric distribution. **A Level**
- To find the probability of making type I and type II errors.
- To perform a χ^2 test to evaluate a distribution model.
- To use a contingency table to calculate expected frequencies and χ^2 contributions.
- To calculate the number of degrees of freedom, and the critical value or the p-value of the test statistic.
- To perform a test for association.

What this leads to

Ch4 Linear regression and correlation

- Hypothesis testing for correlation.

2.1 Testing the mean of a Poisson distribution

Fluency and skills

You can use the Poisson distribution to model the probability of a number of independent, non-simultaneous, random events occurring over a fixed interval of time if you know the value of λ (lambda), the parameter that gives the average number of events.

Key point

In the absence of any other information, the best estimate of λ is the average rate, taken from a random sample.

If you have an estimate and then take a new sample, a hypothesis test will tell you if the sample could reasonably fit your estimate for λ

The **null hypothesis** is that the parameter takes a fixed value $H_0: \lambda = \lambda_0$

The **alternative hypothesis** is either simply that λ_0 is wrong, $H_1: \lambda \neq \lambda_0$ (a 2-tailed test),

or that λ is lower or higher than λ_0, $H_1: \lambda < \lambda_0$ or $H_1: \lambda > \lambda_0$ (a 1-tailed test).

Example 1

A Poisson distribution is believed to have a parameter of 13.2. A sample is taken to test whether or not this is the case. State the null and alternative hypotheses for this test.

The null hypothesis is $H_0: \lambda = 13.2$ and the alternative hypothesis is $H_1: \lambda \neq 13.2$

As with any hypothesis test, you need a **test statistic** in order to evaluate the validity of your hypothesis.

Key point

The **test statistic** when testing the parameter of a Poisson distribution is the total number of events that occur.

Let $X \sim \text{Po}(\lambda)$ where X is the number of events in an interval of length T. A sample of size n is gathered over an interval of length nT and the number of events in the sample, Y, follows a Poisson distribution with parameter $n\lambda$: $Y \sim \text{Po}(n\lambda)$

The total number of events in the sample is compared with the expected number under the null hypothesis. The Poisson distribution only allows integer numbers of outcomes, which is why you cannot compare the average number of events to λ

If this test statistic is sufficiently different from $n\lambda$ then you have sufficient evidence to reject the null hypothesis. You can compare the test result to a critical value, or you can compare the p-value associated with the test statistic to the significance level.

Example 2

A Poisson distribution is believed to have parameter 8.6, $X \sim \text{Po}(8.6)$.
The hypotheses $H_0 : \lambda = 8.6$ and $H_1 : \lambda \neq 8.6$ are tested at the 10% significance level.

a Calculate the critical values for a sample of size 5

A sample of size 5 is taken which has an average rate of 10.4

b State, with a reason whether the null hypothesis is accepted or rejected.

a Let Y be the number of occurrences in a time span 5
Under H_0, $Y \sim \text{Po}(43)$

> $43 = 5 \times 8.6$ (sample size $\times$ average rate)

$P(Y \leq 32) = 4.97\%$ and $P(Y \leq 33) = 6.93\%$

> As this parameter is too large to appear in your tables, test values on your calculator.

$P(Y \geq 54) = 5.86\%$ and $P(Y \geq 55) = 4.39\%$

So the critical values are 32 and 55

> The probability in the tails must be $\leq 10\%$ altogether.

b $5 \times 10.4 = 52$ lies between the critical values so the null hypothesis is accepted.

> Recall that the critical region is bounded by the critical values.

Example 3

A Poisson distribution is believed to have parameter 6.1, $X \sim \text{Po}(6.1)$.

The hypotheses $H_0 : \lambda = 6.1$ and $H_1 : \lambda > 6.1$ are tested at the 5% significance level.

A sample is taken with results 3, 4, 4, 6, 7, 8, 9, 10, 13 and 14.

a Calculate the total number of events, y

b Consider $Y \sim \text{Po}(61)$, the Poisson distribution for the number of events in a sample of size 10 under the null hypothesis. Calculate the probability of obtaining a result at least as extreme as that observed if the parameter is truly 6.1 (i.e. find $P(Y \geq y)$, the p-value for the test statistic).

c State, with a reason whether the null hypothesis is accepted or rejected.

a $3 + 4 + 4 + 6 + 7 + 8 + 9 + 10 + 13 + 14 = 78$

b $P(Y \geq 78) = 0.02032$

> Find the p-value.

c Reject the null hypothesis since the p-value is less than the significance level.

There are two types of possible error you should consider when hypothesis testing.

Key point

A **type I error** is when a null hypothesis which is true is rejected.

A **type II error** is when a null hypothesis which is false is accepted.

The probability of making a type I error is the true level of significance of the hypothesis test.

Example 4

A Poisson distribution is believed to have parameter 1.3, $X \sim \text{Po}(1.3)$.

The hypotheses $H_0: \lambda = 1.3$ and $H_1: \lambda > 1.3$ are tested at the 5% significance level.

A sample of size 8 is taken.

a Find the critical value.

b Find the probability of making a type I error.

a $Y \sim \text{Po}(10.4)$ so $P(Y \geq 16) = 0.06405$ and $P(Y \geq 17) = 0.03681$

Therefore, the critical value is 17.

b $P(\text{type I error}) = P(Y \geq 17) = 0.03681$

Since $0.06405 > 0.05$ and $0.05 > 0.03681$

The sample size, n, is 8 so the new variable, Y, has expected value $8 \times \lambda = 8 \times 1.3 = 10.4$

As Y can only take integer values there is no Y value associated with a probability of exactly 5%.

A type I error occurs when a result at least as extreme as the critical value is found but the true value of λ is 1.3

Exercise 2.1A Fluency and skills

1 A Poisson distribution with assumed parameter 5.31 is being tested to see if this is an overestimate.

   a State the null and alternative hypotheses.

   A sample of size 12 is taken to test at the 10% level.

   b Find any critical values.

2 A Poisson distribution is believed to have parameter 3.4, $X \sim \text{Po}(3.4)$. The hypotheses $H_0: \lambda = 3.4$ and $H_1: \lambda \neq 3.4$ are tested at the 10% level.

   a Calculate the critical values for a sample of size 9

   A sample of size 9 is taken which has an average rate of 3

   b State, with a reason, whether the null hypothesis is accepted or rejected.

3 A Poisson-variable is tested to see if 8.3 is a good estimate for the parameter.

   a State the null and alternative hypotheses.

   A sample of size 9 is taken to test at the 10% level. A total of 87 events occur in the whole sample.

   b Calculate the p-value of this result.

   c State, with a reason, whether the null hypothesis is accepted or rejected.

4 A Poisson distribution is believed to have parameter 14.2, $X \sim \text{Po}(14.2)$. The hypotheses $H_0: \lambda = 14.2$ and $H_1: \lambda > 14.2$ are tested at the 5% level. A sample is taken with results 7, 8, 10, 11, 14, 14, 15, 23, 27, 34.

a Calculate the total number of occurrences, n

b Consider $Y \sim \text{Po}(142)$, the Poisson distribution for the number of events in a sample of size 10. Calculate the probability of obtaining a result at least this extreme if the parameter is truly 14.2 (i.e. find $P(Y \geq n)$, the p-value for the test statistic).

c State, with a reason, whether the null hypothesis is accepted or rejected.

5 A Poisson distribution with assumed parameter 3.9 is being considered. The hypotheses $H_0: \lambda = 3.9$ and $H_1: \lambda > 3.9$ are tested at the 5% level using a sample of size 15

a Calculate any critical values.

b State the probability of making a type I error.

6 A Poisson distribution is being considered. The hypotheses $H_0: \lambda = 6.3$ and $H_1: \lambda < 6.3$ are tested at the 5% level using a sample of size 20. A total of 108 events occur in the whole sample.

a Calculate the p-value of this result.

b State, with a reason, whether the null hypothesis is accepted or rejected.

c Calculate the probability of making a type I error.

Reasoning and problem-solving

Strategy

To solve problems involving hypothesis testing

1. Conduct a test.
2. Find any critical values.
3. Accept or reject the null hypothesis.
4. Interpret the result in context.

Type I and II errors are not exclusive to Poisson distributions. The definitions apply to any hypothesis test.

Example 5

This is a binomial distribution.

A dice is rolled 27 times to test the hypotheses $H_0: p = \frac{1}{6}$ and $H_1: p < \frac{1}{6}$ at the 5% level, where p is the probability of rolling a six. The critical value is found to be 1 six. Calculate the probability of making a type I error.

Let X be the total number of sixes obtained in n rolls.

If $X \sim B\left(27, \frac{1}{6}\right)$ then $P(X \leq 1) = 0.04659$

(2) Identify the critical value using your calculator.

The probability of making a type I error is 0.04659

(4) Interpret the result in context.

Example 6

The number of goals scored per football match is thought to be well modelled by a Poisson distribution with parameter 2.8. Due to recent developments in defensive tactics it is believed that the average number of goals conceded has dropped.

The hypotheses $H_0: \lambda = 2.8$ and $H_1: \lambda < 2.8$ are tested at the 5% level.

A random sample of 24 matches is taken.

a Find any critical values.

The sample contains 49 goals in total.

b State, with a reason, whether the null hypothesis is accepted or rejected. Determine the conclusion to the hypothesis test in context.

a The critical value is 53 as $P(X \leq 53) = 4.34\%$ and $P(X \leq 54) = 5.69\%$.

b Since 49 is less than the critical value, there is sufficient evidence to reject the null hypothesis.

There are, on average, fewer goals conceded per match now.

2 Find the critical values.

3 Reject the null hypothesis.

4 Interpret the result in context.

Exercise 2.1B Reasoning and problem-solving

1 The number of phone calls a call centre receives in a 10-minute period can be modelled by a Poisson distribution with parameter $\lambda = 6.8$. The hypotheses $H_0: \lambda = 6.8$ and $H_1: \lambda \neq 6.8$ are being tested at the 5% level and a sample of seven 10-minute periods are taken.

   a Find any critical values.

   The seven periods contain a total of 59 calls.

   b State, with a reason, whether the null hypothesis is accepted or rejected. Determine the conclusion to the hypothesis test in context.

   c Define a type I error and when one occurs in a hypothesis test.

   d Determine the probability of making a type I error.

2 The number of points scored per game by a basketball team is believed to be well modelled by a Poisson distribution with parameter 102.3. Due to recent training, it is believed that the average number of points scored has increased. The hypotheses $H_0: \lambda = 102.3$ and $H_1: \lambda > 102.3$ are being tested at the 5% level. A random sample of 20 matches is taken.

   a Find any critical values.

   The sample contains a total of 2098 goals.

   b State, with a reason, whether the null hypothesis is accepted or rejected. Determine the conclusion to the hypothesis test in context.

3 The number of accidents per month at a sports centre can be modelled by a Poisson distribution with parameter $\lambda = 0.8$ but new safety procedures have been put in place. The hypotheses $H_0 : \lambda = 0.8$ and $H_1 : \lambda < 0.8$ are being tested at the 5% level. A sample of one year is used and there were five accidents.

a Determine the p-value of the result.

b State, with a reason whether the null hypothesis is accepted or rejected. Determine the conclusion to the hypothesis test in context.

4 The number of customers coming into a book shop in a half-hour period can be modelled by a Poisson distribution with parameter $\lambda = 15$. The hypotheses $H_0 : \lambda = 15$ and $H_1 : \lambda > 15$ are being tested at the 5% level and a sample of twelve 30-minute periods are taken. Over those periods a total of 204 customers came in.

a Find the p-value of this outcome.

b State, with a reason, whether the null hypothesis is accepted or rejected. Determine the conclusion to the hypothesis test in context.

5 The number of light bulbs in a shopping centre which burn out each month can be well modelled by a Poisson distribution with parameter $\lambda = 28$. The hypotheses $H_0 : \lambda = 28$ and $H_1 : \lambda \neq 28$ are being tested at the 10% level. A sample of four random months in a period of several years is taken and a total of 99 light bulbs were replaced in that time.

a Find the p-value of this outcome.

b State, with a reason, whether the null hypothesis is accepted or rejected. Determine the conclusion to the hypothesis test in context.

6 A coin is suspected of being biased. The hypotheses $H_0 : p = \frac{1}{2}$ and $H_1 : p \neq \frac{1}{2}$ are tested at the 5% level, where p is the probability of flipping a tail. The test is performed at the 5% level and the coin is flipped 16 times.

a Find any critical values.

b Determine the probability of making a type I error.

7 The probability of success for a binomial distribution is tested. It is hypothesised that the true value is 0.45 and a sample of size 12 is taken.

a Determine the probability of making a type I error if the significance level is

i 10% ii 5% iii 1%

b Find the probability of making a type II error in each case if the true value is actually 0.6.

2.2 Testing the mean of a geometric distribution

Full A Level section

Fluency and skills

Full A Level

A geometric distribution is used to model situations where a series of independent trials take place. The trials stop as soon as the first success is obtained. The distribution has a single parameter, p, the probability of success.

Key point

The **geometric distribution** is $P(X=x)=(1-p)^{x-1}p$ for $x=1,2,3,..$, where p is the probability of a success and X is the number of trials up to and including the first success. The distribution is written $X \sim \text{Geo}(p)$

Example 1

A fair coin is flipped until a head is obtained.

a Calculate the probability that the head occurs on the fourth flip.

b Calculate the probability that it takes two or more flips for a head to occur.

a $X \sim Geo(0.5)$, $P(X=4)=(1-0.5)^3 \times 0.5=\frac{1}{16}$

For four trials to occur there must have been three failures.

b $P(X \geq 2)=1-P(X=1)=1-\frac{1}{2}=\frac{1}{2}$

This is equal to 1 – P(head appears on first flip)

The **mean** of a geometric distribution is $\frac{1}{p}$ and the **variance** is $\frac{1-p}{p^2}$. You can test the parameter of a geometric distribution to see if a given value is a reasonable estimate of the parameter. By repeating trials until the first success, you can see if the result is outside the likely range.

Example 2

For a geometric distribution, the hypotheses $H_0: p = 0.25$ and $H_1: p < 0.25$ are tested at the 5% significance level.

a Find the critical region.

b Find the expected number of trials before the first success occurs and the variance.

c The first success occurs on the 10th trial. Determine the conclusion to the hypothesis test.

a The critical region is >11 because $P(X \geq 11) = 5.63\% > 5\%$ and $P(X \geq 12) = 4.22\% < 5\%$. So 12 trials is the first value to generate a probability of less than 5%.

These have been calculated using the cumulative distribution function.

b The expected number is $\frac{1}{0.25} = 4$ trials. The variance is $\frac{1-0.25}{0.25^2} = 12$ trials².

c 10 is outside the critical region. There is insufficient evidence to reject the null hypothesis. You conclude that 0.25 is a reasonable estimate for the probability of a success.

Alternatively, logarithms could be used. You want $P(X > k) = 0.05 = (1 - 0.25)^k = 0.75^k$, so $\log 0.05 = k \log 0.75$, giving $k = \frac{\log 0.05}{\log 0.75} = 10.4$. Hence the least integer not smaller than this is 11 so the critical region is >11.

Example 3

A 6-sided dice is rolled repeatedly until a 1 is rolled.

a State the expected number of rolls before a 1 is rolled.

The hypotheses $H_0: p = \frac{1}{6}$ and $H_1: p \neq \frac{1}{6}$ are tested at the 10% significance level, where p is the probability of rolling a 1. The 16th roll is when the first 1 occurs.

b Find the p-value of the result.

c Determine the conclusion of the test.

a $\frac{1}{\frac{1}{6}} = 6$

b $P(X \geq 16) = 1 - P(X \leq 15) = \left(1 - \frac{1}{6}\right)^{15} = 6.49\%$

c As the test is two-tailed and 6.49% > 5%, there is insufficient evidence to reject the null hypothesis.

Exercise 2.2A Fluency and skills

Full A Level

1 A fair dice is rolled until a 6 is obtained. The number of rolls required is modelled by the random variable X

a Calculate the expected number of rolls it takes to obtain a 6

b Calculate the probability it takes longer than that number of rolls to obtain a 6

c Calculate the variance of X

2 A geometric distribution $X \sim \text{Geo}(p)$ is being tested. For each value of q given, use trial and improvement or logarithms to calculate the critical region for the hypotheses $H_0: p = q$ and $H_1: p < q$ at the 5% significance level.

a $q = 0.1$ b $q = 0.05$ c $q = 0.01$

3 A geometric distribution $X \sim \text{Geo}(p)$ is being tested. For the hypotheses $H_0: p = 0.1$ and $H_1: p < 0.1$, find the p-values of these test statistics.

a 25 b 50 c 75

4 For a geometric distribution $X \sim \text{Geo}(p)$, the hypotheses $H_0: p = 0.4$ and $H_1: p < 0.4$ are tested at the 10% significance level.

a Find the critical region.

b Find the expected number of trials required to obtain the first success and its variance.

c The first success occurs on the 6th trial. Determine the conclusion to the hypothesis test.

5 A 12-sided dice is rolled repeatedly until a 12 is obtained.

a Calculate the expected number of rolls before a 12 is obtained.

The hypotheses $H_0: p = \frac{1}{12}$ and $H_1: p \neq \frac{1}{12}$ are tested at the 5% significance level. The 32nd roll gives the first 12 to occur.

b Find the p-value of the result.

c Determine the conclusion of the test.

6 The hypotheses $H_0: p = 0.01$ and $H_1: p > 0.01$ are tested at the 5% significance level, where p is the probability of a random integer generator producing a 50 when it generates a number between 1 and 100. Find the critical region.

Reasoning and problem-solving

Full A Level

Strategy

When testing the mean of a geometric distribution

1. Make sure a geometric distribution is a good model for the situation.
2. Decide the hypotheses being tested in terms of the probability of a success.
3. Find the critical region or calculate the p-value of the test statistic.
4. Determine the conclusion to the test in context.

Example 4

The probability that rain occurs on any random day in a town is thought to be 0.38. A student records the number of days until it next rains.

a The student models this using the geometric distribution. State the hypotheses they are considering at the 5% level.

Actually it takes 11 days to rain.

b Calculate the p-value of the result and determine the conclusion to the test.

c Why might the geometric distribution not necessarily be a good model for the distribution of the number of days until it next rains?

a $H_0: p = 0.38$ and $H_1: p \neq 0.38$, where p is the probability of rain on any given day.

b $P(X \geq 11) = 0.84\% < 2.5\%$ so the result is significant. There is evidence to suggest that the probability of rain is not 0.38.

c Rain occurring on successive days are not necessarily independent events.

(2) Decide the hypotheses being tested in terms of the probability of a success.

(3)(4) Calculate the p-value and determine the conclusion to the test in context.

(1) Make sure a geometric distribution is a good model for the situation.

Example 5

For $X \sim \text{Geo}(p)$ the hypotheses $H_0: p = 0.1$ and $H_1: p > 0.1$ are tested at the 5% level. Explain why this test is pointless.

The critical region does not exist. Since $P(X \leq 1) = 10\% > 5\%$ there is no result which would be sufficient to reject the null hypothesis.

Exercise 2.2B Reasoning and problem-solving

Full A Level

1 A student believes that the probability she can guess the number a person is thinking of is 0.2, but her friend thinks it's less than that. The student guesses different people's numbers until she guesses correctly to test this idea.

a Why might the geometric distribution not necessarily be a good model for this distribution?

b Assuming the student uses the geometric distribution anyway, clearly state the hypotheses she is considering at the 5% level.

Actually it takes 21 guesses before the student guesses correctly.

c Calculate the p-value of the result and determine the conclusion to the test.

2 A gambler has a pack of 52 cards but does not know if it is a regular pack or a trick deck. In a regular pack four of the cards are Aces, so the gambler decides to shuffle the deck and pick a card at random to see if it is an Ace. If it is not then he replaces it and repeats the process, if it is an Ace then he stops.

a Explain why the number of times the gambler picks a card out can be modelled by a geometric distribution.

The gambler had to repeat the process 40 times before he found an Ace.

b Perform a hypothesis test at the 5% level and say if the result suggests that this is a regular pack.

3 For $X \sim \text{Geo}(p)$ the hypotheses $H_0: p = 0.03$ and $H_1: p > 0.03$ and are tested at the 2.5% level.

a Explain why this test cannot produce an outcome.

b Determine the greatest value of p for which a test could produce an outcome at the 2.5% level.

4 A prospector spends his holiday digging for gold. He tries a new location each day and believes that the probability of finding gold in each location is the same. His holiday is 16 days long and he will stop once he has found something.

a Explain why the number of days he takes to find something can be modelled by a Geometric distribution up to the 16th day.

b Determine the smallest probability of success that he could test at the 5% significance level in the time available.

5 A geometric distribution is being tested with the hypotheses $H_0: p = q$ and $H_1: p \neq q$, for some value of q

a Explain why q must be smaller than the significance level of the test if the critical region is to have two tails.

b What happens to the difference of the two critical values as q decreases?

2.3 Errors in hypothesis testing

Full A Level section

Fluency and skills

Full A Level

In hypothesis testing, the null hypothesis is always assumed to be true when determining the critical region. The probability of a result being in the critical region should be less than or equal to the significance level of the test.

Key point

A **type I error** is when a null hypothesis is **rejected** when it is in fact **true**.

A **type II error** is when a null hypothesis is **accepted** when it is in fact **false**.

	Null hypothesis is **accepted**	Null hypothesis is **rejected**
Null hypothesis is **true**	Good outcome	Type I error, probability α
Null hypothesis is **false**	Type II error, probability β	Good outcome

Example 1

It is believed that the probability of success for a geometric distribution is 0.3. A test is performed which leads to the conclusion that this is not the case.

In truth, the probability of a success *is* 0.3. State the name of the error made.

A true null hypothesis has been rejected, so this is a type I error.

The probability of making a type I error is equal to the probability of a result being in the critical region. It is commonly denoted α and is called the **size** of the test.

Key point

The size of a hypothesis test, α, is the probability that a true null hypothesis is rejected.

The probability of a type II error is more difficult to calculate as it depends on the true value of the population parameter being tested. If this was known already then a test would be pointless. An estimation of the probability of making a type II error can be found by using some trial candidates for the true value. The probability of making a type II error is commonly denoted β.

The probabilities of making a type I and a type II error are generally not independent; decreasing one is likely to increase the other. The probability of making a type I error is much easier to control using the significance level.

Key point

The **power** of a hypothesis test is the probability that a false null hypothesis is rejected. It is equal to $1-\beta$

A good hypothesis test has small size and large power.

Example 2

For a geometric distribution $X \sim \text{Geo}(p)$, the hypotheses $H_0: p = 0.01$ and $H_1: p > 0.01$ are tested at the 5% significance level.

a Find the critical region.

b Calculate the size of the test.

a $P(X \le 5) = 4.90\% < 5\%$ and $P(X \le 6) = 5.85\% > 5\%$

The critical region is $X \le 5$

b Size of test $= P(X \text{ in critical region} | H_0) = P(X \le 5) = 4.90\%$.

Calculate the probabilities under H_0: $X \sim \text{Geo}(0.01)$ using the geometric distribution function on your calculator.

The size of the test is the probability under H_0 of obtaining a result in the critical region.

Example 3

The mean of a Normal distribution with standard deviation 2.2 is being tested at the 10% level. The null hypothesis is $\mu = -0.8$ and the alternative hypothesis is $\mu \ne -0.8$. The true mean is $\mu = -3.2$.

a Find the probability of a type II error.

b Calculate the power of the test.

a For $X \sim N(\mu, 2.2^2)$

P(type II error) = P(H_0 accepted)

= P(X not in critical region)

Critical region:

For $X \sim N(-0.8, 2.2^2)$,

$P(X < a) = 0.05 \Rightarrow a = -4.42$

$P(X > b) = 0.05 \Rightarrow P(X < b) = 0.95 \Rightarrow b = 2.82$

For this 2-tailed test the critical region is $X < -4.42$ and $X > 2.82$.

For $\mu = -3.2$,

$P(-4.42 < X < 2.82) = 0.707$

So if the true mean is -3.2, the probability of obtaining a value outside of the critical region is 70.7%.

b The power is $1 - 0.707 = 0.293$

Use the null hypothesis to define the critical region for the test.

Use your calculator to find inverse Normal probabilities.

Exercise 2.3A Fluency and skills

Full A Level

1 For a geometric distribution $X \sim \text{Geo}(p)$, the hypotheses $H_0: p = 0.32$ and $H_1: p < 0.32$ are tested at the 10% significance level.

a Find the critical region.

b Calculate the size of the test.

The first success occurs on the 15th trial.

c Determine the conclusion to the hypothesis test.

2 With reference to the correct hypotheses, explain the difference between a type I and a type II error.

3 A Normal distribution with standard deviation 6.25 is being tested at the 5% significance level. The null hypothesis is $H_0: \mu = 15.7$ and the alternative is $H_1: \mu < 15.7$

a State the probability of a type I error.

b Find the probability of a type II error if actually $\mu = 5.7$

c Find the probability of a type II error if actually $\mu = 25.7$

4 A Poisson distribution is tested at the 5% level with hypotheses $H_0: \lambda = 6.8$ and $H_1: \lambda > 6.8$

a Find the critical region.

b Calculate the probability of a type I error.

The true value is $\lambda = 9.7$

c Calculate the probability of a type II error.

d Calculate the power.

5 For a binomial distribution $X \sim B(n, p)$ the hypotheses $H_0: p = 0.2$ and $H_1: p \neq 0.2$ are tested at the 10% level. 20 trials are performed and the critical region is $X = 0$ or $X > 7$. Calculate the probability of a type II error and the power when the true value of p is

a 0.3 b 0.4 c 0.6

6 A Normal distribution is being tested at the 5% level with hypotheses $H_0: \mu = 0$ and $H_1: \mu > 0$. The standard deviation is 4.5

a Calculate the power if the true mean is 12.1

b Explain why the same power is found if the true mean is −12.1

c Explain why the power tends to 1 as the true mean tends to infinity.

Reasoning and problem-solving

Full A Level

Strategy

To solve problems involving type I and type II errors

1. If necessary, decide on the significance level for the test.
2. Calculate the critical value.
3. Calculate the probability of making a type I or type II error.
4. Answer the question in context.

Example 4

A Normal distribution with standard deviation 4 is being tested with hypotheses $H_0: \mu = 7.7$ and $H_1: \mu < 7.7$. Calculate the critical value and hence the probability of making a type II error if the true mean is 2.5 when the significance level is

a 10% **b** 5% **c** 1%

a For significance level 10%, the critical value is 2.57

$P(X > 2.57 \mid \mu = 2.5) = 49.3\%$

b The critical value is 1.12

$P(X > 1.12 \mid \mu = 2.5) = 63.4\%$

c The critical value is −1.61

$P(X > -1.61 \mid \mu = 2.5) = 84.8\%$

(2) Find the critical value.

Cumulative probabilities for the Normal distribution can be found using your calculator.

(3) Calculate the probability of a type II error.

Consider a Poisson distribution being tested with hypotheses $H_0: \lambda = 10$ and $H_1: \lambda > 10$. At the 5% significance level the critical region is $X \geq 16$. The table shows the power for a range of true parameter values.

$X \sim Po(\lambda)$

True value of λ	10	11	12	13	14	15	16	17	18
P(*X* *not* in critical region) $= P(X \leq 15) = \beta$	0.951	0.907	0.844	0.764	0.669	0.568	0.466	0.371	0.287
Power of test $= 1 - \beta$	0.049	0.093	0.156	0.236	0.331	0.432	0.534	0.629	0.713

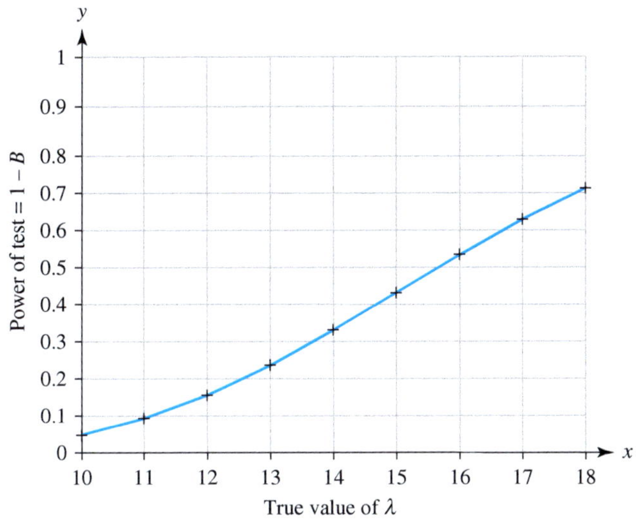

The power is less than 0.5 for values of λ less than 15.7, exemplifying that if the true value of the parameter is much smaller than the critical value of the test then there is a better-than-even chance that the false null hypothesis is accepted.

When the true value is much higher than the critical value the power becomes close to 1.

Note that the power isn't technically defined when the null hypothesis really is the true value.

Example 5

A new medical test for detecting a rare illness is being trialled. The null hypothesis is that a person being tested does not have the illness.

a Explain why the probability of a type II error should be made as small as possible.

b This makes the probability of a type I error larger. Why might this not be a problem?

a A type II error is when a false null hypothesis is accepted. A medical test should aim to not overlook somebody who genuinely has the illness.

Answer the question in context. (4)

b A type I error is when a true null hypothesis is rejected. If the test claims that a person has the illness when they don't, then a repeat of the test will help decide.

Answer the question in context. (4)

Exercise 2.3B Reasoning and problem-solving

Full A Level

1 A Normal distribution with standard deviation 0.6 is being tested with the hypotheses $H_0: \mu = 0.3$ and $H_1: \mu > 0.3$. Calculate the critical value, the probability of making a type II error if the true mean is 1.3 and the power when the significance level is

a 10%

b 5%

c 1%

2 A Poisson distribution is being tested with hypotheses $H_0: \lambda = 1.7$ and $H_1: \lambda \neq 1.7$. The true value of the mean is 2.3. For the significance level of

a 10%

b 5%

Find

i the critical region

ii the probability of a type II error

iii the power.

3 For a binomial distribution the hypotheses $H_0: p = \frac{1}{3}$ and $H_1: p \neq \frac{1}{3}$ are tested at the 2% level. Twenty trials are performed and the critical region is $X < 2$ or $X > 12$. The true value of p is 0.4

a State how many successes you would expect to see (using the true value of p).

b Calculate the probability of a type II error and the power.

Instead of 20 trials, 2000 trials are performed. The new critical region is $X < 618$ or $X > 716$

c State how many successes you would expect to see.

d Giving a reason, explain how you would expect the power of this test to compare with the 20-trial test.

4 A treasure hunter is configuring their metal detector to find buried valuables.

a Explain why the probability of a type II error should be made as small as possible.

b This makes the probability of a type I error larger. Why might this not be a problem?

5 The hypotheses $H_0: \lambda = 4.25$ and $H_1: \lambda < 4.25$ are tested at the 5% level, where λ is the parameter of a population Poisson distribution.

a Find the critical region.

b Find the probability of a type I error.

The true mean is 9.25

c Calculate the probability of a type II error and the power of the test.

6 A binomially distributed population is tested with the hypotheses $H_0: p = 0.2$ and $H_1: p > 0.2$ at the 5% level. Different-sized samples are taken to test this. The true value of p is 0.35. Calculate the critical region and hence the power when the sample size is

a 10 b 20 c 30

7 The number of goals scored by a team playing away from home in a football match can be modelled by a Poisson distribution with parameter $\lambda = 1.2$. A team has undergone a training regime that it is hoped will increase the number of goals they score away. The next five away games are played and the hypotheses $H_0: \lambda = 6$ and $H_1: \lambda > 6$ are tested at the 5% significance level.

a Explain why the hypotheses take this form.

b Find the critical region and calculate the probability of making a type I error.

Post-training, the number of goals scored away can now be modelled by a Poisson distribution with parameter $\lambda = 1.4$

c Calculate the probability of making a type II error.

d Explain why this error wouldn't be very concerning to a fan of the team.

e Explain why this error would be very concerning to a sports-betting company, who gamble on the results of matches.

2.4 Goodness of fit

Fluency and skills

A dice is rolled 300 times and the results in the table are obtained.

Outcome	1	2	3	4	5	6
Observed frequency	44	59	35	57	54	51

Does this data look like it has come from an unbiased dice?

With an unbiased dice, you expect each outcome to occur 50 times, although in practice this is unlikely to happen.

So how different does the outcome need to be before you believe the dice to be biased at the 5% significance level?

In a hypothesis test for how well a proposed model describes a situation, the **null hypothesis** is that the model fits the data well and the **alternative hypothesis** is that it does not.

If the sample you take is unlikely to have come from the proposed model then you have evidence to suggest that the proposed model was not a good fit.

Key point

The χ^2 **(chi squared) statistic** gives a way to measure how different the observed values are from the expected values.

The χ^2 statistic, X, is a calculation involving the observed frequencies $\{O_i\}$ and the expected frequencies $\{E_i\}$

$$X = \sum_{i=1}^{n} \frac{(O_i - E_i)^2}{E_i}$$

For the dice, $X = \frac{(44-50)^2 + (59-50)^2 + (35-50)^2 + (57-50)^2 + (54-50)^2 + (51-50)^2}{50} = 8.16$

You can compare this particular statistic to the χ^2 **distribution** with five **degrees of freedom**.

The χ^2 distribution with k degrees of freedom is the distribution of a sum of the squares of k independent standard Normal random variables.

Key point

If $Y_1, Y_2, \ldots, Y_k$ are independent random variables and $Y_i \sim N(0, 1)$ for $i = 1, 2, \ldots, k$ then $(Y_1^2 + Y_2^2 + \ldots + Y_k^2)$ follows a χ^2 distribution with k degrees of freedom.

Degrees of freedom are independent parameters. Since the expected frequencies must sum to the same as the observed frequencies, knowing all but one determines the last one.

The shape of a χ^2 distribution depends on its parameter, k. For the purposes of this section, it is useful to use the cumulative distribution $F_k(x)$ where

$F_k(x) = P(X \le x)$ where $X \sim \chi_k^2$

The larger the value of k, the higher the expected value of X. In fact $E(X) = k$.

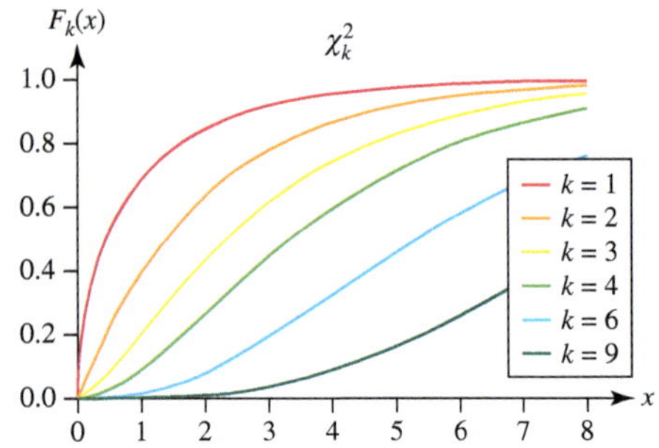

The χ^2 distribution can be used in goodness of fit tests for random variables, like comparing the number of times each possible outcome is obtained when a dice is rolled 300 times with the expected results.

In the dice example, if a dice is rolled 300 times and with six possible outcomes then there are five degrees of freedom. Once you know the frequencies of five of the outcomes, the sixth is given by (300 - sum of known frequencies), so it is no longer considered free.

Key point

For a random variable with n outcomes, the number of degrees of freedom is n with 1 subtracted for each parameter estimated using the data.

If the p-value of the χ^2 statistic is smaller than the significance level, or if the χ^2 statistic is larger than the critical value found using the χ^2 distribution, then there is sufficient evidence to reject the null hypothesis.

In the dice example, the probability of a χ^2 statistic greater than or equal to 8.16 is $0.1476 > 0.05$, so this is *not* an unusually high statistic and you accept the null hypothesis that the dice is unbiased.

Calculator

Try it on your calculator

You can use a calculator to find p-values for a χ^2 distribution.

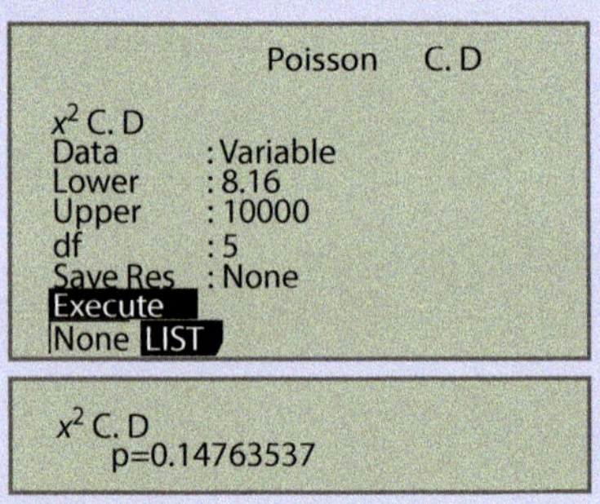

Activity

Find out how to calculate $P(X \ge 8.16)$ where $X \sim \chi_5^2$ on *your* calculator.

You can use a statistical table to find critical values. As the dice example involves five degrees of freedom, and the test is conducted at a $5\% = 0.05$ significance level, look at the percentage points table for the χ^2 distribution and find the value corresponding to $v = 5$ and 0.05

v	0.995	0.990	0.975	0.950	0.900	0.100	0.050
1	0.000	0.000	0.001	0.004	0.016	2.705	3.841
2	0.010	0.020	0.051	0.103	0.211	4.605	5.991
3	0.072	0.115	0.216	0.352	0.584	6.251	7.815
4	0.207	0.297	0.484	0.711	1.064	7.779	9.488
5	0.412	0.554	0.831	1.145	1.610	9.236	**11.070**

This shows that $P(X \geq 11.070) = 0.05$, so 11.070 is the critical value for this test. When testing a statistic in a hypothesis test you should look at the probability of getting a result equal to or more extreme than the observed result. In the case of a χ^2 statistic, 'more extreme' always means bigger.

$8.16 < 11.070$ so this result is not significant at the 5% level. You accept the null hypothesis and say the dice is not biased.

Key point

For this test to work accurately, every expected frequency needs to be greater than 5. Note that observed frequencies may be 5 or lower.

Example 1

A random variable is believed to follow the distribution given in the table. To test this at the 10% level, a sample of size 80 is taken, with the observed frequencies also given in the table.

x	1	2	3	4
$P(X=x)$	0.225	0.275	0.4425	0.0575
Observed frequency	24	16	28	12

a Calculate the expected frequencies if the distribution is a good model for the random variable.

b Calculate the contributions to the χ^2 statistic.

c State the number of degrees of freedom and hence calculate the critical value.

d Determine whether the null hypothesis, H_0 = the random variable follows the distribution given in the table, is accepted or rejected.

e A new sample of size 100 is taken, which has a χ^2 statistic of 6.5. Calculate the p-value for this statistic and determine whether the null hypothesis would be accepted or rejected at the 10% level.

a $x = 1, 0.225 \times 80 = 18$

$x = 2, 0.275 \times 80 = 22$

$x = 3, 0.4425 \times 80 = 35.4$

$x = 4, 0.0575 \times 80 = 4.6$

b

x	1	2	3 to 4
$P(X=x)$	0.225	0.275	0.5
Expected frequency	$0.225 \times 80 = \mathbf{18}$	$0.275 \times 80 = \mathbf{22}$	$0.5 \times 80 = \mathbf{40}$
Observed frequency	24	16	40
χ^2 contributions	$\frac{(24-18)^2}{18} = \mathbf{2}$	$\frac{(16-22)^2}{22} = \mathbf{1.636}$	$\frac{(40-40)^2}{40} = \mathbf{0}$

As the expected frequency of 4 is <5, group 4 with 3 before proceeding.

c $4 - 1 - 1 = 2$ degrees of freedom.

$v = 2$, $10\% = 0.1$, critical value 4.605

One degree of freedom is lost because the number of categories is reduced by 1. Another is lost because the expected frequencies must sum to 80 to match the observed frequencies.

(*Continued on the next page*)

d The total χ^2 statistic is

$2 + 1.636 + 0 = 3.636, 3.636 < 4.605$

The χ^2 statistic is smaller than the critical value so accept the null hypothesis; the distribution given in the table is a good fit.

The critical value is the value z for which $P(X \geq z) = 0.1$

e $x = 4$, Expected frequency $= 0.0575 \times 100 = 5.75 > 5$

For this sample there are no expected frequencies smaller than 5 so there are 4 outcomes.

Degrees of freedom $= 4 - 1 = 3$

$X \sim \chi^2_3$

$P(X \geq 6.5) \approx P(6.5 \leq X \leq 1000)$

$= 0.08966...$

Your calculator might require both a lower and upper bound. If so, use an arbitrarily high upper bound.

$0.08966... < 10\%$ so reject the null hypothesis

The probability of getting the observed result or something more extreme is smaller than the significance level.

Exercise 2.4A Fluency and skills

1 A 10-sided dice is rolled 80 times to see if it can be described as unbiased.

  a Calculate the expected frequency of each outcome.

  b Calculate the number of degrees of freedom and hence the critical value at the 5% significance level.

The results obtained are given in this table.

x	1	2	3	4	5	6	7	8	9	10
Observed frequency	14	5	9	2	11	6	8	7	5	13

  c Perform a χ^2 test at the 5% level to determine if the dice could be described as unbiased.

2 Calculate the p-value of each of these χ^2 statistics at the 5% significance level.

  a 12.45 for 5 degrees of freedom

  b 12.45 for 8 degrees of freedom

  c 17.9 for 8 degrees of freedom

  d 17.9 for 11 degrees of freedom.

3 For each of these distributions,

  a Calculate the number of degrees of freedom and hence the critical value at the 10% level.

  b Perform a χ^2 test at the 10% level to see if the proposed expected frequencies are good models.

  i

x	1	2	3	4	5
Expected frequency	50	40	30	40	50
Observed frequency	35	45	50	45	35

ii

x	1	2	3	4	5	6
Expected frequency	31	24	19	15	12	10
Observed frequency	25	25	25	12	12	12

iii

x	1	2	3	4	5	6	7
Expected frequency	4	6	10	12	9	7	3
Observed frequency	7	5	14	9	11	5	0

4 A variable is known to be well modelled by a binomial distribution, but the exact value of the parameter is yet to be determined. One person conjectures it to be 0.35, but another believes it to be 0.39. They each take 1000 samples, each with 5 trials. The results and their expected frequencies are given in the table correct to 1 decimal place.

Number of successes in 5 trials	0	1	2	3	4	5
Expected for $p = 0.35$	116.0	312.4	336.4	181.1	48.8	5.3
Expected for $p = 0.39$	84.5	270.0	345.2	220.71	70.6	9.0
Observed frequency	99	292	342	201	59	7

a Using the values in the table, calculate the χ^2 statistic for the $p = 0.35$ row.

b Using the values in the table, calculate the χ^2 statistic for the $p = 0.39$ row.

The critical value is 11.1 for 5 degrees of freedom.

c Determine each person's conclusion.

d Explain why there is no contradiction between the two conclusions.

Reasoning and problem-solving

Strategy

To decide if a model is a good description of a situation

1. Use the probabilities in the model to determine the expected frequencies in the sample. You may have to estimate some of these using the observed frequencies in the sample.
2. Group outcomes when the expected frequency is less than 5 so that the χ^2 statistic is relevant.
3. Calculate the χ^2 statistic and compare to the critical value for the number of degrees of freedom.
4. Determine the conclusion to the hypothesis test and decide if the model is a good description.

Example 2

A situation is tested to see if the binomial distribution can be used to model it.
A total of 100 samples of size 8 are taken. Recorded in the table are the number of successes obtained across the 100 samples.

Number of successes in 8 trials	0	1	2	3	4	5	6	7	8
Observed frequency	20	32	25	15	7	1	0	0	0

a Calculate the value of p, the probability of an individual success.

b For your value of p, calculate the expected frequencies for 100 samples.

c Group the outcomes to make the χ^2 test valid.

d Calculate the number of degrees of freedom and hence the critical value at the 5% level.

e Calculate the χ^2 statistic for the sample.

f Determine the conclusion to the goodness of fit test.

a $0\times20+1\times32+2\times25+\ldots+8\times0=160$ successes across the 800 tests

$p=\frac{160}{800}=0.2$

Divide the total number of successes by the total number of trials to get the relative frequency.

b

Number of successes in 8 trials (x)	0	1	2
Probability of x successes	${}^8C_0(0.2)^0(0.8)^8=0.1678$	${}^8C_1(0.2)^1(0.8)^7=0.3355$	${}^8C_2(0.2)^2(0.8)^6=0.2936$
Expected frequency	16.78	33.55	29.36

Number of successes in 8 trials (x)	3	4	5
Probability of x successes	${}^8C_3(0.2)^3(0.8)^5=0.1468$	${}^8C_4(0.2)^4(0.8)^4=0.0459$	${}^8C_5(0.2)^5(0.8)^3=0.00918$
Expected frequency	14.68	4.59	0.918

Number of successes in 8 trials (x)	6	7	8
Probability of x successes	${}^8C_6(0.2)^6(0.8)^2=0.00115$	${}^8C_7(0.2)^7(0.8)^1=0.0000819$	${}^8C_8(0.2)^8(0.8)^0=0.00000256$
Expected frequency	0.115	0.00819	0.000256

Use the binomial formula ${}^nC_r p^r(1-p)^{n-r}$ to calculate probabilities.

1 Determine the expected frequencies.

(*Continued on the next page*)

c

Number of successes in 8 trials	0	1	2	3	4–8
Expected frequency	16.78	33.55	29.36	14.68	5.63
Observed frequency	20	32	25	15	8

d There are $5 - 1 - 1 = 3$ degrees of freedom.

For χ_3^2 at the 5% level the critical value is 7.815

e χ^2 statistic $= \frac{(20-16.78)^2}{16.78} + \frac{(32-33.55)^2}{33.55} + \frac{(25-29.36)^2}{29.36} + \frac{(15-14.68)^2}{14.68} + \frac{(8-5.63)^2}{5.63}$

$= 2.35$

f $2.35 < 7.815$. There is insufficient evidence to reject the null hypothesis that the binomial distribution is a good model for the situation.

(2) The outcomes have been grouped because the χ^2 distribution requires all expected frequencies to be at least 5

There are now 5 groups, but the expected frequencies must sum to 100 and the parameter p has been estimated using the data.

(3) Compare the χ^2 statistic to the critical value.

(4) Determine the conclusion to the hypothesis test.

Note that the conclusion doesn't mention $p = 0.2$. It's unlikely this number is exactly correct for the situation. You only conclude that some binomial distribution is a good model.

Example 3

The number of people waiting in a queue is being tested to see if the Poisson distribution can be used to model how many of them are served in under 2 minutes. A sample of 75 trials is taken and the following outcomes are observed.

Number of people served in under 2 minutes	0	1	2	3	4	5	6	7+
Observed frequency	7	12	10	11	19	12	4	0

a Calculate the mean number of outcomes per trial, λ

b For your value of λ, calculate the expected frequencies for 75 trials.

c Group outcomes to make the χ^2 test valid.

d Calculate the number of degrees of freedom and hence the critical value at the 5% level.

e Calculate the χ^2 statistic for the sample.

f Determine the conclusion to the goodness of fit test.

a $\lambda = ((0 \times 7) + (1 \times 12) + (2 \times 10) + (3 \times 11) + (4 \times 19) + (5 \times 12) + (6 \times 4) + (7 \times 0)) \div 75 = 3$

b

Number of people served in under 2 minutes (x)	0	1	2	3
Probability of x people being served in an interval	$\frac{3^0 e^{-3}}{0!} = 0.0498...$	$\frac{3^1 e^{-3}}{1!} = 0.1494...$	$\frac{3^2 e^{-3}}{2!} = 0.2240...$	$\frac{3^3 e^{-3}}{3!} = 0.2240...$
Expected frequency	3.73	11.20	16.80	16.80

(1) Determine the expected frequencies by multiplying the probability by the number of samples, 75

(*Continued on the next page*)

Number of people served in under 2 minutes (x)	4	5	6	7+
Probability of x people being served in an interval	$\frac{3^4 e^{-3}}{4!}$ $=0.1680\ldots$	$\frac{3^5 e^{-3}}{5!}$ $=0.1008\ldots$	$\frac{3^6 e^{-3}}{6!}$ $=0.0504\ldots$	$1-(0.0498\ldots+$ $\ldots+0.0504\ldots)=$ 0.0336
Expected frequency	12.60	7.56	3.78	2.52

c

Number of people served in under 2 minutes	0–1	2	3	4	5	6+
Expected frequency	14.93	16.80	16.80	12.60	7.56	6.30
Observed frequency	19	10	11	19	12	4

(2) Group the outcomes so that all frequencies are at least 5

d There are $6-1-1=4$ degrees of freedom.

For χ_4^2 at the 5% level the critical value is 9.488

There are now 6 groups, but the expected frequencies must sum to 75 and the value of λ has been calculated using the observed frequencies.

e $\chi^2 \text{ statistic} = \frac{(19-14.93)^2}{14.93} + \frac{(10-16.80)^2}{16.80} + \frac{(11-16.80)^2}{16.80}$

$+\frac{(19-12.60)^2}{12.60} + \frac{(12-7.56)^2}{7.56} + \frac{(4-6.30)^2}{6.30}$

$=12.54$

(3) Calculate the χ^2 statistic.

f $12.54 > 9.488$, the critical value is smaller than the test statistic, so there is sufficient evidence to reject the null hypothesis.

(3) Compare the χ^2 statistic to the critical value.

The Poisson distribution is not a good model for the number of people served in less than 2 minutes.

(4) Determine the conclusion to the hypothesis test.

If any Poisson distribution were going to be a good fit, it would be the one that had the exact value of λ that was calculated from the sample.

Exercise 2.4B Reasoning and problem-solving

1 A situation is believed to be well modelled by a binomial distribution with unknown parameter p. 75 samples of size 4 are taken and the data is given in this table.

Number of successes out of 4	0	1	2	3	4
Observed frequency	5	13	19	23	15

a Calculate the value of p, the probability of an individual success.

b For your value of p, calculate the expected frequencies for 75 samples.

c Group outcomes until no group's expected frequency is less than 5

d Calculate the χ^2 statistic for the sample.

e Calculate the number of degrees of freedom and hence the critical value at the 5% level.

f Determine the conclusion to the goodness of fit test.

2 A situation is tested to see if the Poisson distribution can be used to model it. A sample of 50 trials gave these results.

Outcome	0	1	2	3	4	5+
Observed frequency	4	16	12	12	6	0

a Calculate the mean number of outcomes per trial, λ, and calculate the expected frequencies for 50 trials.

b Perform a goodness of fit test at the 5% significance level.

3 The discrete uniform distribution, where every outcome is equally likely, is tested at the 10% significance level to see if it is a good model for a situation. The results shown in the table are obtained. Perform a goodness of fit test at the 10% significance level.

Outcome	4	7	22	25	41
Observed frequency	8	15	9	23	10

4 The number of attempts, r, needed to pass a driving test can be modelled as
$\mathrm{P}(X=r)=p(1-p)^{r-1}, r \geq 1$

A group of 150 people are surveyed and these results are obtained.

r	1	2	3	4	5	6+
Observed frequency	84	47	15	3	1	0

a Calculate the average number of attempts, $\bar{x}$, needed to pass the driving test.

It can be shown that the best estimate of p from a sample is $\frac{1}{\bar{x}}$

b Calculate p and hence the expected frequencies.

c Perform a goodness of fit test at the 5% significance level.

5 A variable is believed to follow a geometric distribution with parameter 0.3. A sample of size 70 is taken and the results in the table are obtained.

x	1	2	3	4	>4
Observed frequency	25	19	14	5	7

a Calculate the expected frequency of each outcome.

b Perform a goodness of fit test at the 10% significance level.

Full A Level

2.5 Contingency tables

Fluency and skills

You have seen that correlation describes linear relationships between variables, but correlation cannot describe relationships between variables that do not have a continuous scale. There can be many different relationships between variables. A relationship is known as an **association**.

In the table is data about regional preferences for types of fruit. A sample of 267 people from the north, east and south of a city were asked which fruit they liked most, apples, oranges, or bananas. The **observed frequencies** are shown in the **contingency table** below.

Observed frequencies	North	East	South	Totals
Apples	32	9	44	85
Oranges	41	53	20	114
Bananas	8	32	28	68
Totals	81	94	92	267

It certainly looks like there might be regional preferences but how can you be sure? The first stage is to work out what the table would look like if the variables, region and type of fruit, were independent. You would expect $267 \times \frac{81}{267} \times \frac{85}{267}$ or 25.8 (to 3 sf) people to be from the north with apples as their favourite.

If the variables are independent then the probability of liking apples and coming from the north is P(likes apples) × P(from the north)

Using this method, the table below shows all the expected frequencies (values given to 3 sf).

Expected frequencies	North	East	South
Apples	25.8	29.9	29.3
Oranges	34.6	40.1	39.3
Bananas	20.6	23.9	23.4

The next stage is to set up a hypothesis to test. The **null hypothesis** being considered is H_0: there is no association between the variables. The **alternative hypothesis** is H_1: there is some association. This example is to be tested at the 5% significance level.

Note that you don't test for positive or negative association, just the presence of any association.

The individual χ^2 contributions are shown in the table.

χ^2 contributions $\frac{(O_i - E_i)^2}{E_i}$	**North**	**East**	**South**
Apples	1.49	14.61	7.38
Oranges	1.18	4.15	9.48
Bananas	7.71	2.75	0.904

This sum of the individual contributions gives a value of $X = 49.7$ (all values to 3 sf).

The total number of people is known so, if you know the number of people from the north and the east, the number of people from the south is fixed. The region contributes $3 - 1 = 2$ to the total degrees of freedom. Similarly, if you know the number of people who prefer apples and oranges then the number of people who prefer bananas is fixed at 267 – (number of people who prefer apples or oranges). The fruit preference contributes $3 - 1 = 2$ to the total degrees of freedom.

Overall this example has $(3 - 1)(3 - 1) = 4$ degrees of freedom.

Key point

For an $m \times n$ contingency table you should use the χ^2 distribution with $(m - 1) \times (n - 1)$ degrees of freedom.

The final stage is to decide if you have sufficient evidence to reject the null hypothesis. To do this, you compare $X = 49.7$ to the distribution with four degrees of freedom.

If the χ^2 statistic is larger than the critical value found using the χ^2 distribution, or if the p-value is less than the significance level, then you have sufficient evidence to reject the null hypothesis.

For the fruit preferences example being considered, the χ^2 statistic of 49.7 compares with a critical value of 9.49 at the 5% level for χ^2_4. Since the calculated value of X is larger than the critical value, there is sufficient evidence to reject the null hypothesis. It is likely there is an association between the regions and fruit preferences given.

For $X \sim \chi^2_4$, $P(X > 9.49) = 0.05$, you can find this value in the Percentage points of the χ^2 distribution table.

Example 1

A test is performed to see if there is any association between age and movie genre preferences. The contingency table shows the recorded data.

	Action	Comedy	Romance	Totals
<25 years	41	34	27	102
25–50 years	35	38	32	105
>50 years	7	12	19	38
Totals	83	84	78	245

a Draw a table of expected frequencies.

b Draw a table of χ^2 contributions and calculate the χ^2 test statistic.

c Calculate the number of degrees of freedom and hence the p-value of the test statistic.

d Determine the conclusion to the test at the 2.5% significance level.

a

E_i	Action	Comedy	Romance
<25 years	$245\times\frac{83}{245}\times\frac{102}{245}$ $=34.6$	$245\times\frac{84}{245}\times\frac{102}{245}$ $=35.0$	$245\times\frac{78}{245}\times\frac{102}{245}$ $=32.5$
25–50 years	$245\times\frac{83}{245}\times\frac{105}{245}$ $=35.6$	$245\times\frac{84}{245}\times\frac{105}{245}$ $=36.0$	$245\times\frac{78}{245}\times\frac{105}{245}$ $=33.4$
>50 years	$245\times\frac{83}{245}\times\frac{38}{245}$ $=12.9$	$245\times\frac{84}{245}\times\frac{38}{245}$ $=13.0$	$245\times\frac{78}{245}\times\frac{38}{245}$ $=12.1$

83 out of 245 people prefer action movies and 102 out of 245 people are <25

b

	Action	Comedy	Romance
<25 years	1.20	0.03	0.93
25–50 years	0.01	0.11	0.06
>50 years	2.70	0.08	3.94

For example $\frac{(41-34.6)^2}{34.6}=1.20$ is the χ^2 contribution for Action and <25 years.

χ^2 statistic $=1.20+0.03+0.93+0.01+0.11+0.06$
$+2.70+0.08+3.94=9.06$ (3 sf)

c The 3×3 table has $(3-1)\times(3-1)=4$ degrees of freedom.
For $X\sim\chi^2_4$, $P(X\geq9.06)=5.96\%$
The p-value is 5.96%.

Use your calculator to find the probability.

d $5.96\%>2.5\%$. The p-value is larger than the significance level, so there is insufficient evidence to reject the null hypothesis.

The method changes slightly when you have a 2×2 contingency table or if any of the observed frequencies is less than 5. The χ^2 statistic is calculated differently. This is called **Yates' correction.**

Key point

$$X_{\text{Yates}} = \sum_{i=1}^{N} \frac{\left(|O_i - E_i| - 0.5\right)^2}{E_i}$$

Example 2

The lunch choices of two year groups at a school are tested for association. A sample of 63 students in Year 12 and Year 13 are asked whether they prefer Fish 'n' Chips or Salad.

O_i	Fish 'n' Chips	Salad	Totals
Year 12	19	12	31
Year 13	14	18	32
Totals	33	30	63

a Draw a table of expected frequencies (correct to 3 sf).

b Draw a table of contributions and calculate the χ^2 test statistic.

c At the 10% level the critical value is 2.71. Determine the conclusion to the test.

a

E_i	Fish 'n' Chips	Salad
Year 12	16.2	14.8
Year 13	16.8	15.2

b

E_i	Fish 'n' Chips	Salad
Year 12	0.327	0.357
Year 13	0.315	0.348

The χ^2 contributions (3 sf) have been calculated using Yates' correction.

The test statistic is 1.35 (3 sf).

This is the sum of the individual contributions.

c The test statistic is less than the critical value. There is insufficient evidence to reject the null hypothesis.

Exercise 2.5A Fluency and skills

For each of these contingency tables:

a Draw a table of expected frequencies.

b Draw a table of χ^2 contributions and calculate the χ^2 test statistic.

c Calculate the number of degrees of freedom and hence the critical value at the 5% level.

d Calculate the p-value of the χ^2 statistic.

e Determine the conclusion to the test.

1

	A	*B*	*C*	Totals
X	57	92	5	154
Y	7	32	1	40
Z	75	100	23	198
Totals	139	224	29	392

2

	A	*B*	*C*	*D*	Totals
W	16	13	16	10	55
X	8	13	8	17	46
Y	17	7	17	17	58
Z	8	15	18	18	59
Totals	49	48	59	62	218

3

	A	*B*	*C*	Totals
X	79	57	85	221
Y	61	72	62	195
Totals	140	129	147	416

4 (Remember to use Yates' correction.)

	A	*B*	Totals
X	70	124	194
Y	50	45	95
Totals	120	169	289

5 A test is performed to see if there is any association between the colour a hen is and how many eggs it lays per week. The contingency table shows the recorded data.

O_i	0–1	2–3	4+	Totals
Brown	6	13	16	35
White	17	20	19	56
Totals	23	33	35	91

a Draw a table of expected frequencies.

b Draw a table of χ^2 contributions and calculate the χ^2 test statistic.

c Calculate the number of degrees of freedom and hence the p-value of the test statistic.

d Determine the conclusion to the test at the 5% level.

6 A snack shop tests meal deal choices for association. A sample of sales is checked for sandwich and drink choices.

O_i	Chicken	BLT	Ham & Cheese	Egg Mayo	Totals
Orange	31	28	19	6	84
Apple	19	29	17	8	73
Water	24	21	20	9	74
Totals	74	78	56	23	231

a Draw a table of expected frequencies.

b Draw a table of χ^2 contributions and calculate the χ^2 test statistic.

c At the 10% level the critical value is 10.6. Determine the conclusion to the test.

Reasoning and problem-solving

Strategy

To solve problems involving possible association

1. Perform a test for association.
2. Identify positive or negative association by comparing observed and expected values for large χ^2 contributions.
3. Identify where the observed and expected frequencies closely match from small χ^2 contributions.

Example 3

An airline takes a survey of passengers of various heights to gauge views on the amount of legroom available.

	>1.80 m	<1.50 m	Between 1.50 m and 1.80 m	Totals
Too little	67	32	304	403
Too much	19	48	415	482
About right	48	51	379	478
Totals	134	131	1098	1363

a Perform a test for association at the 5% level.

b Comment on the association.

(Continued on the next page)

a

	>1.80m	<1.50m	Between 1.50m and 1.80m
Too little	39.6	38.7	324.6
Too much	47.4	46.3	388.3
About right	47.0	45.9	385.1

This is the table of expected frequencies.

	>1.80m	<1.50m	Between 1.50m and 1.80m
Too little	18.96	1.16	1.31
Too much	17.02	0.06	1.84
About right	0.02	0.57	0.096

This is the table of χ^2 contributions.

The χ^2 statistic is 41.0 (3 sf)

The critical value at the 5% level is = 9.49

so there is sufficient evidence to reject the null hypothesis.

(1) Perform a test for association.

b There is association between height and opinions on how much legroom is available.

(2) Identify positive association.

The large contributions from people over 1.80 m show that their opinions differ most from what is expected. Looking at the observed and expected frequencies there is evidence for a strong positive association between being tall and believing there is too little legroom. Similarly, there is a strong negative association between being tall and believing there is too much legroom.

(2) Identify positive or negative association by comparing observed and expected values for large χ^2 contributions.

The number of short people who believe there is too much legroom is very close to what is expected.

(3) Identify where the observed and expected frequencies closely match from small χ^2 contributions.

Exercise 2.5B Reasoning and problem-solving

In questions **1** to **3**:

a Perform a test for association at the 5% level.

b Comment on the association.

1 A group of people is surveyed to see if there is any association between their eye colour and their favourite colour.

Observed	Blue eyes	Green eyes	Brown eyes	Totals
Blue	99	37	38	174
Red	34	29	33	96
Yellow	32	34	27	93
Totals	165	100	98	363

2 In a coastal town, the colours of rocks are compared with whether they were found on the beach or by the river.

Observed	Grey	Brown	Totals
Beach	76	64	140
River	107	59	166
Totals	183	123	306

3 Live-action role players are asked which ancient tribes they most enjoy playing and which weapon they most like to carry.

Observed	Axe	Sword	Bow	Spear	Totals
Celts	10	29	20	34	93
Vikings	12	24	10	31	77
Huns	0	14	31	24	69
Totals	22	67	61	89	239

4 A restaurant is looking for association between choices of starter and main course from its menu.

Observed	Pizza	Pasta	Totals
Soup	87	121	208
Salad	104	96	200
Totals	191	217	408

a Perform a test for association at the 10% level.

b Comment on the association.

5 A toy company tests its new range of coloured monster toys for association between child gender and their preference of monster colour.

Observed	Red	Blue	Yellow	Totals
Boys	75	71	57	203
Girls	58	79	60	197
Totals	133	150	117	400

a Perform a test for association at the 1% level.

b Comment on the association.

6 For a 2 × 2 contingency table Yates' correction can be written as

$$X_{\text{Yates}} = \frac{N\left(|ad-bc|-\frac{N}{2}\right)^2}{T_X T_Y T_A T_B}$$

for tables in the form

Observed	A	B	Totals
X	a	b	T_X
Y	c	d	T_Y
Totals	T_A	T_B	N

For every 2 × 2 contingency table in this section, check that this formula agrees with your value.

2 Summary and review

Chapter summary

- A Poisson distribution has a parameter, λ, which is the average rate that the event occurs.
- When you don't know the value of λ you can estimate it as $\lambda = \lambda_0$ and then perform a hypothesis test:
  - The null hypothesis is $H_0: \lambda = \lambda_0$
  - The alternative hypothesis is either:
    - $H_1: \lambda \neq \lambda_0$, a 2-tailed test
    - $H_1: \lambda < \lambda_0$, a 1-tailed test, or
    - $H_1: \lambda > \lambda_0$, a 1-tailed test.
- If the null hypothesis is true, then the distribution of the number of events in a sample of size n is a Poisson distribution with parameter λn. If the number of events in the sample is sufficiently unlikely then you reject the null hypothesis.
- Sometimes a null hypothesis which is true is rejected due to an unlikely result. This is called a type I error.
- The test gives evidence for or against λ_0 being a reasonable estimate of the parameter of the distribution.
- A geometric distribution has a parameter, p, which is the probability of a success occurring on each trial in a sequence of trials. The sequence stops once a success occurs. **Full A Level**
- When you don't know the value of p you can estimate it as $p = p_0$ and then perform a hypothesis test:
  - The null hypothesis is $H_0: p = p_0$
  - The alternative hypothesis is either:
    - $H_1: p \neq p_0$, a 2-tailed test
    - $H_1: p < p_0$, a 1-tailed test, or
    - $H_1: p > p_0$, a 1-tailed test.
- A type I error is when a null hypothesis which is true is rejected. The probability of making one is called the size of the test and is denoted α.
- A type II error is when a null hypothesis which is false is accepted. The probability of making one is denoted β and the power of the test is $1 - \beta$.
- A contingency table shows the observed frequencies, $\{O_i\}$, from a sample which splits data by two discrete variables.
- In a test for association, the null hypothesis is that there is no association between the variables and the alternative hypothesis is that there is some association.
- You should calculate a table of expected frequencies, $\{E_i\}$, to show what the sample would look like if the two variables were independent.
- The χ^2 test statistic, X, can be calculated from the observed and expected frequencies.

$$X = \sum_i \frac{(O_i - E_i)^2}{E_i}$$

- For a 2×2 contingency table, or when an observed frequency is less than 5, the test statistic is calculated using Yates' correction.

$$X_{\text{Yates}} = \sum \frac{\left(|O_i - E_i| - 0.5\right)^2}{E_i}$$

- The χ^2 distribution with k degrees of freedom is the distribution of a sum of the squares of k independent standard Normal random variables.
- The χ^2 test statistic can be compared to the appropriate χ^2 distribution to test for significance. For an $m \times n$ contingency table you should use the χ^2 distribution with $(m-1)\times(n-1)$ degrees of freedom.
- If the statistic is larger than the critical value, or if the p-value is less than the significance level, then the null hypothesis H_0 is rejected.
- Large χ^2contributions highlight association and you should compare observed and expected frequencies so that you can comment on the association.

Check and review

You should now be able to...	Review Questions
✔ Test the parameter of a Poisson distribution.	1, 2, 3
✔ Test the parameter of a geometric distribution. Full A Level	4, 5
✔ Calculate the probability of making type I and type II errors.	2, 4
✔ Perform a χ^2 test to see if a distribution is a reasonable model for a situation.	6, 7, 9, 10
✔ Use a contingency table to calculate expected frequencies and χ^2 contributions.	6, 7, 8, 9, 12
✔ Calculate the number of degrees of freedom, and the critical value or the p-value of the test statistic.	6, 7, 10, 11
✔ Perform a test for association.	11, 12

1 A Poisson distribution is believed to have parameter 1.9 and a test is performed to see if it is higher than this.

a State the null and alternative hypotheses.

A sample of size 19 is taken to test at the 5% level. A total of 48 events occurs in the whole sample.

b Calculate the p-value of this result.

c State, with a reason whether the null hypothesis is accepted or rejected.

2 It is thought that $X \sim \text{Po}(0.75)$. The hypotheses $H_0: \lambda = 0.75$ and $H_1: \lambda < 0.75$ are tested at the 5% level. A sample of size 29 is taken.

a Find the critical value.

b Find the probability of making a type I error.

Full A Level

3 The number of road accidents on a country road per month is well modelled by a Poisson distribution with parameter 1.3. It is hoped that new traffic-calming measures will reduce this. The hypotheses $H_0 : \lambda = 1.3$ and $H_1 : \lambda < 1.3$ are being tested at the 10% level. The next 8 months form a sample.

a Find any critical values.

There are seven accidents in those 8 months.

b State, with a reason whether the null hypothesis is accepted or rejected. Determine the conclusion to the hypothesis test in context.

Full A Level

4 For a geometric distribution, the hypotheses $H_0 : p = 0.042$ and $H_1 : p < 0.042$ are tested at the 5% significance level.

a Find the critical region.

b Calculate the probability of making a type II error if the true value of p is 0.024.

5 An 8-sided dice is rolled repeatedly until an 8 is obtained.

a State the expected number of rolls before an 8 is obtained.

The hypotheses $H_0 : p = \frac{1}{8}$ and $H_1 : p \neq \frac{1}{8}$ are tested at the 10% significance level, where p is the probability of rolling an 8. The 30th roll gives the first 8 to occur.

b Find the p-value of the result.

c Determine the conclusion of the test.

6 A population is being tested to see if the Poisson distribution can be used to model it. A sample of 80 trials is taken and these outcomes are observed.

Outcome	0	1	2	3	4+
Observed frequency	34	13	24	9	0

a Calculate the mean number of outcomes per trial, λ, and calculate the expected frequencies for 80 trials.

b Perform a goodness of fit test at the 5% significance level.

7 A situation is believed to be well modelled by a binomial distribution with unknown parameter p. 250 samples of size 6 are taken and the data is given in this table.

Outcome	0	1	2	3	4	5	6
Observed frequency	0	10	32	54	76	63	15

a Calculate the value of p, the probability of an individual success.

b For your value of p, calculate the expected frequencies for 250 samples.

c Group outcomes until no group's expected frequency is less than 5

d Calculate the number of degrees of freedom and hence the critical value at the 5% level.

e Calculate the χ^2 statistic for the sample.

f Determine the conclusion to the goodness of fit test.

8 For this contingency table

O_i	*A*	*B*	*C*	Totals
X	103	71	35	209
Y	96	84	41	221
Z	111	58	49	218
Totals	310	213	125	648

a Draw a table of expected frequencies,

b Draw a table of χ^2 contributions and calculate the χ^2 test statistic.

9 Given this contingency table and its table of expected values, calculate the χ^2 test statistic using Yates' correction.

O_i	*A*	*B*	Totals
X	24	36	60
Y	60	24	84
Totals	84	60	144

E_i	*A*	*B*
X	35	25
Y	49	35

10 For this contingency table, calculate

O_i	*A*	*B*	*C*	*D*	*E*	Totals
X	29	37	25	87	41	219
Y	14	39	93	20	53	219
Z	10	5	2	7	28	52
Totals	53	81	120	114	122	490

a The number of degrees of freedom,

b The critical value at the 5% significance level.

11 The heights and birthplaces of UK adults are tested for association at the 5% level. A large sample is taken with 24 degrees of freedom. The χ^2 statistic is found to be 34.91. Calculate the *p*-value of this statistic and determine the conclusion to the test in context.

12 A car designer is testing public opinion on four new models (*A*, *B*, *C*, *D*) for association with preference for colour. Using the contingency table

O_i	*A*	*B*	*C*	*D*	Totals
Red	100	84	60	80	324
Blue	55	64	57	78	254
Silver	30	36	27	23	116
Totals	185	184	144	181	694

a Draw a table of expected frequencies,

b Draw a table of χ^2 contributions and calculate the χ^2 test statistic.

There are 6 degrees of freedom so the critical value at the 5% level is 12.59

c Determine the conclusion to the test and comment on the association.

Exploration

Going beyond the exams

Did you know?

You have studied type I and type II errors but you can also have type III errors. There are a number of interpretations for this type of error that you might like to look up. One popular interpretation is when you correctly reject a null hypothesis but for the wrong reasons!

Research

A confusion matrix is a type of contingency table that reports on the number of false-positives, false-negatives, true-positives and true-negatives.

For example, imagine a piece of software that is being used to identify whether 16 photographs were of males or females. It could achieve the following results:

		Actual class	
		Male	Female
Predicted class	Male	5	6
	Female	2	3

The software achieved 5 true-positives, 6 false-positives, 2 false-negatives and 3 true-negatives.

You should research how this information can be used to calculate the F_1 score – which measures the accuracy of a test.

Research

There are many tests for goodness of fit that you could research. These include

- the Kolmogorov-Smirnov test
- Kuiper's test
- the Shapiro-Wilk test

Do some research and find some examples of what you might use each test for.

ICT

You can use Excel to find p-values if you know the chi-square value and the degrees of freedom of a problem.

For example, a toy company claims that 30% of their football stickers are team badges, 60% are players and 10% are managers.

Would a random sample of 100 stickers made up of 20 team badges, 75 players and 5 managers be consistent with the company's claim (at a 0.05 significance level)?

You should set up the problem in full to confirm the degrees of freedom as 2 and that 9.58 is the chi-square test statistic.

To find the probability that the chi-square is greater than 9.58, you can then use Excel to input the following:

=1 – CHISQ.DIST(9.58,2,TRUE)

Where TRUE indicates that you want to use a cumulative distribution.

Using this result from Excel, determine whether you would accept the null hypothesis.

2 Assessment

1 Using traditional thread, a machine produces parachute material with an average of 3.2 flaws per 5 m². It is claimed that a new type of thread will reduce the number of flaws. In a random sample of 10 m² using the new thread, 2 flaws are found.

a State the null and alternative hypotheses to test this claim. **[1]**

b Perform the test at a significance level of 5% **[5]**

c Explain what is meant by a type I error and find the probability that a type I error is *not* made in the test in part **b**. Full A Level **[2]**

2 A business receives on average 12 telephone calls per hour throughout the day. The day after a newspaper advertising campaign, it receives 90 calls between 10 am and 4 pm.

a Using a significance level of 5%, test the hypothesis that the campaign has increased the rate of telephone calls. **[7]**

b State the probability of a type I error for your test and explain what this means. Full A Level **[2]**

(You may assume that the number of calls in a given period follows a Poisson distribution.)

3 In the manufacture of precision microscope lenses it is known that over a long period 23% of the total production is rejected at the final quality control stage. Immediately following changes to the production process, the first rejection was of the eighth lens produced. Based on this observation the company wishes to test whether the changes have altered the proportion of lenses rejected. The test should be performed at the 5% significance level.

Let X be a random variable for the number of lenses produced up to and including the first rejection.

a Give the distribution of X, stating any assumptions made, and state the null and alternative hypotheses for this test. **[3]**

b Find, under the null hypothesis, $P(X \geq 8)$ **[2]**

c Use your answer to part **b** to state the conclusion of your test. **[2]**

4 A Geiger counter indicates radioactive decay by a series of clicks heard through a loudspeaker, one click per decay. For a particular sample of radioactive material, the number of decays over a randomly chosen 10 second interval is measured. The source of the activity is thought to be a material with a mean number of decays of 2.7 per 10 second period. A scientist wishes to test, at a significance level of 2.5%, this hypothesis against the alternative hypothesis that the source is actually a material with a greater mean. You should assume that decays occur randomly in time.

a State the null and alternative hypotheses for this test. **[2]**

b Find the critical region for this test. **[3]**

c Given that the true value of the mean is 5, find the probability of a type II error and the power of this test. Full A Level **[5]**

5 At a major accident blackspot, records over a long period show that road traffic accidents occur randomly with a mean of 1.7 per month. During 36 months after traffic calming measures were put in place, the data shown in the table was recorded for X, the number of accidents per month.

x	0	1	2	3	4	>4
f(x)	8	12	10	5	1	0

a Copy and complete the given table of probabilities and expected values based upon the assumption that the traffic calming measures made no difference. You should justify the distribution you use. [4]

x	0	1	2	3	4	>4
P($X = x$)						
Expected frequency						

b Carry out a goodness of fit test comparing the observed data to the expected frequencies, stating the number of degrees of freedom for the test and giving a clear conclusion on the effectiveness of the new measures. Use a significance level of 5% [7]

6 a A random variable X has a geometric distribution with parameter $p = 0.4$. Copy and complete the table of probabilities for X. [2]

x	1	2	3	4	5	6	>6
P($X = x$)				0.0864	0.05184	0.0311	0.04666

b 45 observations are taken from a geometric population, parameter p. The results are shown in the table.

x	1	2	3	4	5	6	>6
Frequency	13	7	8	6	5	2	4

Using a goodness of fit test and your results in part **a**, test at the 10% significance level whether or not $p = 0.4$ [9]

Full A Level

7 A teacher wishes to assess her students on a recently covered topic. She decides to use two assessment methods, a traditional problem-solving exam (A) and a multiple choice exam (B). The teacher believes that there is no significant difference in exam performance between these two assessment methods.

		Result		
		Fail	Pass	Distinction
Exam	A	7	18	8
	B	7	14	6

a Create a table of expected frequencies on the assumption that she is correct. Use the same column and row headings as the table of observed frequencies and include column and row totals. [2]

b Test the teacher's belief at the 5% significance level. [6]

8 A health researcher assesses whether patient safety is affected by weekend staffing levels. She obtains data for nursing staff drug errors and the day of the week on which they occur. The drug errors are classified as type A: errors with no significant consequences to the patient, type B: errors with minor consequences and type C: errors with serious consequences.

		Time		
		Weekday	Weekend	Total
Error type	A	45	18	63
	B	24	8	32
	C	14	5	19
	Total	83	31	114

Test, at a 5% significance level, whether the timing (weekday or weekend) of a drug error affects its seriousness. [8]

Mathematical formulae

For AS and A Level Further Maths

The following mathematical formulae will be provided for you. Statistics formulae, for this course, are provided at the start of this section. Formulae for the entire A Level course are then also provided for your reference.

Statistics

Discrete distributions

For a discrete random variable X taking values x_i with probabilities $P(X=x_i)$

Expectation (mean): $E(X)=\mu=\Sigma x_i P(X=x_i)$

Variance: $Var(X)=\sigma^2=\Sigma(x_i-\mu)^2 P(X=x_i)=\Sigma x_i^2 P(X=x_i)-\mu^2$

For a function g(X): $E(g(X))=\Sigma g(x_i) P(X=x_i)$

The probability generating function of X is $G_X(t)=E(t^X)$ and

$E(X)=G'_X(1)$ and $Var(X)=G''_X(1)+G'_X(1)-[G'_X(1)]^2$

For $Z=X+Y$, where X and Y are independent: $G_Z(t)=G_X(t)\times G_Y(t)$

Full A Level

Standard discrete distributions:

Distribution of X	$P(X=x)$	Mean	Variance	PGF
Binomial $B(n, p)$	$\binom{n}{x}p^x(1-p)^{n-x}$	np	$np(1-p)$	$(1-p+pt)^n$
Poisson $Po(\lambda)$	$e^{-\lambda}\frac{\lambda^x}{x!}$	λ	λ	$e^{\lambda(t-1)}$
Geometric $Geo(p)$ on 1, 2, ...	$p(1-p)^{x-1}$	$\frac{1}{p}$	$\frac{1-p}{p^2}$	$\frac{pt}{1-(1-p)t}$
Negative binomial on r, $r+1$, ...	$\binom{x-1}{r-1}p^r(1-p)^{x-r}$	$\frac{r}{p}$	$\frac{r(1-p)}{p^2}$	$\left(\frac{pt}{1-(1-p)t}\right)^r$

Full A Level

Continuous distributions

For a continuous random variable X having probability density function f

Expectation (mean): $E(X)=\mu=\int x f(x)\,dx$

Variance: $Var(X)=\sigma^2=\int(x-\mu)^2 f(x)\,dx=\int x^2 f(x)\,dx-\mu^2$

For a function g(X): $E(g(X))=\int g(x) f(x)\,dx$

Cumulative distribution function: $F(x_0)=P(X\le x_0)=\int_{-\infty}^{x_0} f(t)\,dt$

Standard continuous distributions:

Distribution of X	PDF	Mean	Variance
Normal $N(\mu, \sigma^2)$	$\frac{1}{\sigma\sqrt{2\pi}}e^{-\frac{1}{2}\left(\frac{x-\mu}{\sigma}\right)^2}$	μ	σ^2
Uniform (Rectangular) on $[a, b]$	$\frac{1}{b-a}$	$\frac{1}{2}(a+b)$	$\frac{1}{12}(b-a)^2$

Correlation and regression

For a set of n pairs of values (x_i, y_i)

$$S_{xx} = \sum(x_i - \bar{x})^2 = \sum x_i^2 - \frac{\left(\sum x_i\right)^2}{n}$$

$$S_{yy} = \sum(y_i - \bar{y})^2 = \sum y_i^2 - \frac{\left(\sum y_i\right)^2}{n}$$

$$S_{xy} = \sum(x_i - \bar{x})(y_i - \bar{y}) = \sum x_i y_i - \frac{\left(\sum x_i\right)\left(\sum y_i\right)}{n}$$

The product moment correlation coefficient is

$$r = \frac{S_{xy}}{\sqrt{S_{xx}S_{yy}}} = \frac{\sum(x_i - \bar{x})(y_i - \bar{y})}{\sqrt{\left\{\sum(x_i - \bar{x})^2\right\}\left\{\sum(y_i - \bar{y})^2\right\}}} = \frac{\sum x_i y_i - \frac{\left(\sum x_i\right)\left(\sum y_i\right)}{n}}{\sqrt{\left(\sum x_i^2 - \frac{\left(\sum x_i\right)^2}{n}\right)\left(\left(\sum y_i^2\right) - \frac{\left(\sum y_i\right)^2}{n}\right)}}$$

The regression coefficient of y on x is $b = \frac{S_{xy}}{S_{xx}} = \frac{\sum(x_i - \bar{x})(y_i - \bar{y})}{\sum(x_i - \bar{x})^2}$

Least squares regression line of y on x is $y = a + bx$ where $a = \bar{y} - b\bar{x}$

Residual Sum of Squares (RSS) $= S_{yy} - \frac{(S_{xy})^2}{S_{xx}} = S_{yy}(1 - r^2)$

Spearman's rank correlation coefficient is $r_s = 1 - \frac{6\sum d^2}{n(n^2 - 1)}$

Expectation algebra

For independent random variables X and Y

$\mathrm{E}(XY) = \mathrm{E}(X)\mathrm{E}(Y)$ $\quad$ $\mathrm{Var}(aX \pm bY) = a^2\mathrm{Var}(X) + b^2\mathrm{Var}(Y)$

Sampling distributions

(i) Tests for mean when σ is known

For a random sample $X_1, X_2, \ldots, X_n$ of n independent observations from a distribution having mean μ and variance σ^2

$\overline{X}$ is an unbiased estimator of μ, with $\mathrm{var}(\overline{X}) = \dfrac{\sigma^2}{n}$

S^2 is an unbiased estimator of σ^2, where $S^2 = \dfrac{\sum(X_i - \overline{X})^2}{n-1}$

For a random sample of n observations from $\mathrm{N}(\mu, \sigma^2)$, $\dfrac{\overline{X} - \mu}{\sigma/\sqrt{n}} \sim \mathrm{N}(0, 1)$

For a random sample of n_x observations from $\mathrm{N}(\mu_x, \sigma_x^2)$, and, independently, a random sample of n_y observations from $\mathrm{N}(\mu_y, \sigma_y^2)$, $\dfrac{(\overline{X} - \overline{Y}) - (\mu_x - \mu_y)}{\sqrt{\dfrac{\sigma_x^2}{n_x} + \dfrac{\sigma_y^2}{n_y}}} \sim \mathrm{N}(0, 1)$

(ii) Tests for variance and mean when σ is not known

For a random sample of n observations from $\mathrm{N}(\mu, \sigma^2)$

$$\frac{(n-1)S^2}{\sigma^2} \sim X_{n-1}^2$$

$\dfrac{\overline{X} - \mu}{S/\sqrt{n}} \sim t_{n-1}$ (Also valid in matched-pairs situations)

For a random sample of n_x observations from $\mathrm{N}(\mu_x, \sigma_x^2)$ and, independently, a random sample of n_y observations from $\mathrm{N}(\mu_y, \sigma_y^2)$

$$\frac{S_x^2/\sigma_x^2}{S_y^2/\sigma_y^2} \sim F_{n_x-1, n_y-1}$$

If $\sigma_x^2 = \sigma_y^2 = \sigma^2$ (unknown) then

$$\frac{(\overline{X} - \overline{Y}) - (\mu_x - \mu_y)}{\sqrt{S_P^2 \left(\dfrac{1}{n_x} + \dfrac{1}{n_y}\right)}} \sim t_{n_x+n_y-2} \quad \text{where} \quad S_P^2 = \frac{(n_x-1)S_x^2 + (n_y-1)S_y^2}{n_x + n_y - 2}$$

Non parametric tests

Goodness-of-fit test and contingency tables: $\sum \dfrac{(O_i - E_i)^2}{E_i} \sim \chi_v^2$

Pure Mathematics

Summations

$$\sum_{r=1}^{n} r^2 = \frac{1}{6}n(n+1)(2n+1) \qquad \sum_{r=1}^{n} r^3 = \frac{1}{4}n^2(n+1)^2$$

Matrix transformations

Anticlockwise rotation through θ about O: $\begin{pmatrix} \cos\theta & -\sin\theta \\ \sin\theta & \cos\theta \end{pmatrix}$

Reflection in the line $y = (\tan\theta)x$: $\begin{pmatrix} \cos 2\theta & \sin 2\theta \\ \sin 2\theta & -\cos 2\theta \end{pmatrix}$

Area of a sector

$A = \frac{1}{2}\int r^2 \, d\theta$ (polar coordinates)

Complex numbers

$\{r(\cos\theta + i\sin\theta)\}^n = r^n(\cos n\theta + i\sin n\theta)$

The roots of $z^n = 1$ are given by $z = e^{\frac{2\pi k i}{n}}$ for $k = 0, 1, 2, \ldots, n-1$

Maclaurin's and Taylor's series

$$f(x) = f(0) + xf'(0) + \frac{x^2}{2!}f''(0) + \ldots + \frac{x^r}{r!}f^{(r)}(0) + \ldots$$

$e^x = \exp(x) = 1 + x + \frac{x^2}{2!} + \ldots + \frac{x^r}{r!} + \ldots$ for all x

$\ln(1+x) = x - \frac{x^2}{2} + \frac{x^3}{3} - \ldots + (-1)^{r+1}\frac{x^r}{r} + \ldots$ $(-1 < x \le 1)$

$\sin x = x - \frac{x^3}{3!} + \frac{x^5}{5!} - \ldots + (-1)^r \frac{x^{2r+1}}{(2r+1)!} + \ldots$ for all x

$\cos x = 1 - \frac{x^2}{2!} + \frac{x^4}{4!} - \ldots + (-1)^r \frac{x^{2r}}{(2r)!} + \ldots$ for all x

$\arctan x = x - \frac{x^3}{3} + \frac{x^5}{5} - \ldots + (-1)^r \frac{x^{2r+1}}{2r+1} + \ldots$ $(-1 \le x \le 1)$

Vectors

Vector products: $\mathbf{a} \times \mathbf{b} = |\mathbf{a}||\mathbf{b}|\sin\theta\,\hat{\mathbf{n}} = \begin{vmatrix} \mathbf{i} & \mathbf{j} & \mathbf{k} \\ a_1 & a_2 & a_3 \\ b_1 & b_2 & b_3 \end{vmatrix} = \begin{pmatrix} a_2b_3 - a_3b_2 \\ a_3b_1 - a_1b_3 \\ a_1b_2 - a_2b_1 \end{pmatrix}$

$$\mathbf{a}\cdot(\mathbf{b}\times\mathbf{c}) = \begin{vmatrix} a_1 & a_2 & a_3 \\ b_1 & b_2 & b_3 \\ c_1 & c_2 & c_3 \end{vmatrix} = \mathbf{b}\cdot(\mathbf{c}\times\mathbf{a}) = \mathbf{c}\cdot(\mathbf{a}\times\mathbf{b})$$

If A is the point with position vector $\mathbf{a} = a_1\mathbf{i} + a_2\mathbf{j} + a_3\mathbf{k}$ and the direction vector $\mathbf{b}$ is given by $\mathbf{b} = b_1\mathbf{i} + b_2\mathbf{j} + b_3\mathbf{k}$, then the straight line through A with direction vector $\mathbf{b}$ has cartesian equation

$$\frac{x-a_1}{b_1} = \frac{y-a_2}{b_2} = \frac{z-a_3}{b_3} (= \lambda)$$

The plane through A with normal vector $\mathbf{n} = n_1\mathbf{i} + \mathrm{n}_2\mathbf{j} + \mathrm{n}_3\mathbf{k}$ has cartesian equation $n_1x + n_2y + n_3z + d = 0$ where $d = -\mathbf{a.n}$

The plane through non-collinear points A, B and C has vector equation

$\mathbf{r} = \mathbf{a} + \lambda(\mathbf{b} - \mathbf{a}) + \mu(\mathbf{c} - \mathbf{a}) = (1 - \lambda - \mu)\mathbf{a} + \lambda\mathbf{b} + \mu\mathbf{c}$

The plane through the point with position vector $\mathbf{a}$ and parallel to $\mathbf{b}$ and $\mathbf{c}$ has equation

$\mathbf{r} = \mathbf{a} + s\mathbf{b} + t\mathbf{c}$

The perpendicular distance of (α, β, γ) from $n_1x + n_2y + n_3z + d = 0$ is $\dfrac{|n_1\alpha + n_2\beta + n_3\gamma + d|}{\sqrt{n_1^2 + n_2^2 + n_3^2}}$

Hyperbolic functions

$\cosh^2 x - \sinh^2 x = 1$

$\sinh 2x = 2\sinh x\cosh x$

$\cosh 2x = \cosh^2 x + \sinh^2 x$

$\operatorname{arcosh} x = \ln\left\{x + \sqrt{x^2 - 1}\right\}$ $\quad (x \geq 1)$

$\operatorname{arsinh} x = \ln\left\{x + \sqrt{x^2 + 1}\right\}$

$\operatorname{artanh} x = \frac{1}{2}\ln\left(\frac{1+x}{1-x}\right)$ $\quad (|x| < 1)$

Conics

	Ellipse	**Parabola**	**Hyperbola**	**Rectangular Hyperbola**
Standard Form	$\frac{x^2}{a^2}+\frac{y^2}{b^2}=1$	$y^2=4ax$	$\frac{x^2}{a^2}-\frac{y^2}{b^2}=1$	$xy=c^2$
Parametric Form	$(a\cos\theta, b\sin\theta)$	$(at^2, 2at)$	$(a\sec\theta, b\tan\theta)$ $(\pm a\cosh\theta, b\sinh\theta)$	$\left(ct, \frac{c}{t}\right)$
Eccentricity	$e<1$ $b^2=a^2(1-e^2)$	$e=1$	$e>1$ $b^2=a^2(e^2-1)$	$e=\sqrt{2}$
Foci	$(\pm ae, 0)$	$(a, 0)$	$(\pm ae, 0)$	$\left(\pm\sqrt{2}c, \pm\sqrt{2}c\right)$
Directrices	$x=\pm\frac{a}{e}$	$x=-a$	$x=\pm\frac{a}{e}$	$x+y=\pm\sqrt{2}c$
Asymptotes	none	none	$\frac{x}{a}=\pm\frac{y}{b}$	$x=0, y=0$

Differentiation

f(x)	**f′(x)**
$\arcsin x$	$\frac{1}{\sqrt{1-x^2}}$
$\arccos x$	$-\frac{1}{\sqrt{1-x^2}}$
$\arctan x$	$\frac{1}{1+x^2}$
$\sinh x$	$\cosh x$

f(x)	**f′(x)**
$\cosh x$	$\sinh x$
$\tanh x$	$\text{sech}^2 x$
$\text{arsinh}\, x$	$\frac{1}{\sqrt{1+x^2}}$
$\text{arcosh}\, x$	$\frac{1}{\sqrt{x^2-1}}$
$\text{artanh}\, x$	$\frac{1}{1-x^2}$

Integration (+ constant; $a > 0$ where relevant)

$\mathbf{f}(x)$	$\int \mathbf{f}(x)\,\mathrm{d}x$
$\sinh x$	$\cosh x$
$\cosh x$	$\sinh x$
$\tanh x$	$\ln\cosh x$
$\dfrac{1}{\sqrt{a^2-x^2}}$	$\arcsin\left(\dfrac{x}{a}\right) \quad (\lvert x\rvert < a)$
$\dfrac{1}{a^2+x^2}$	$\dfrac{1}{a}\arctan\left(\dfrac{x}{a}\right)$
$\dfrac{1}{\sqrt{x^2-a^2}}$	$\operatorname{arcosh}\left(\dfrac{x}{a}\right), \ln\{x+\sqrt{x^2-a^2}\} \quad (x > a)$
$\dfrac{1}{\sqrt{a^2+x^2}}$	$\operatorname{arsinh}\left(\dfrac{x}{a}\right), \ln\{x+\sqrt{x^2+a^2}\}$
$\dfrac{1}{a^2-x^2}$	$\dfrac{1}{2a}\ln\left\lvert\dfrac{a+x}{a-x}\right\rvert = \dfrac{1}{a}\operatorname{artanh}\left(\dfrac{x}{a}\right) \quad (\lvert x\rvert < a)$
$\dfrac{1}{x^2-a^2}$	$\dfrac{1}{2a}\ln\left\lvert\dfrac{x-a}{x+a}\right\rvert$

Arc length

$$s = \int\sqrt{1+\left(\frac{\mathrm{d}y}{\mathrm{d}x}\right)^2}\,\mathrm{d}x \quad \text{(cartesian coordinates)}$$

$$s = \int\sqrt{\left(\frac{\mathrm{d}x}{\mathrm{d}t}\right)^2+\left(\frac{\mathrm{d}y}{\mathrm{d}t}\right)^2}\,\mathrm{d}t \quad \text{(parametric form)}$$

$$s = \int\sqrt{r^2+\left(\frac{\mathrm{d}r}{\mathrm{d}\theta}\right)^2}\,\mathrm{d}\theta \quad \text{(polar form)}$$

Surface area of revolution

$$s_x = 2\pi\int y\sqrt{\left(1+\left(\frac{\mathrm{d}y}{\mathrm{d}x}\right)^2\right)}\,\mathrm{d}x \quad \text{(cartesian coordinates)}$$

$$s_x = 2\pi\int y\sqrt{\left(\frac{\mathrm{d}x}{\mathrm{d}t}\right)^2+\left(\frac{\mathrm{d}y}{\mathrm{d}t}\right)^2}\,\mathrm{d}t \quad \text{(parametric form)}$$

$$s_x = 2\pi\int r\sin\theta\sqrt{r^2+\left(\frac{\mathrm{d}r}{\mathrm{d}\theta}\right)^2}\,\mathrm{d}\theta \quad \text{(polar form)}$$

Full A Level

Mechanics

Centres of mass

For uniform bodies:

Triangular lamina: $\frac{2}{3}$ along median from vertex

Circular arc, radius r, angle at centre 2α: $\frac{r\sin\alpha}{\alpha}$ from centre

Sector of circle, radius r, angle at centre 2α: $\frac{2r\sin\alpha}{3\alpha}$ from centre

Solid hemisphere, radius r: $\frac{3}{8}r$ from centre

Hemispherical shell, radius r: $\frac{1}{2}r$ from centre

Solid cone or pyramid of height h: $\frac{1}{4}h$ above the base on the line from centre of base to vertex

Conical shell of height h: $\frac{1}{3}h$ above the base on the line from centre of base to vertex

Motion in a circle

Transverse velocity: $v = r\dot{\theta}$

Transverse acceleration: $\dot{v} = r\ddot{\theta}$

Radial acceleration: $-r\dot{\theta}^2 = -\frac{v^2}{r}$

Statistical tables

For AS and A Level Further Maths

The following statistical tables will be provided for you.

Binomial cumulative distribution function

The tabulated value is $P(X \le x)$, where X has a binomial distribution with index n and parameter p

$p=$	0.05	0.10	0.15	0.20	0.25	0.30	0.35	0.40	0.45	0.50
$n=5, x=0$	0.7738	0.5905	0.4437	0.3277	0.2373	0.1681	0.1160	0.0778	0.0503	0.0313
1	0.9774	0.9185	0.8352	0.7373	0.6328	0.5282	0.4284	0.3370	0.2562	0.1875
2	0.9988	0.9914	0.9734	0.9421	0.8965	0.8369	0.7648	0.6826	0.5931	0.5000
3	1.0000	0.9995	0.9978	0.9933	0.9844	0.9692	0.9460	0.9130	0.8688	0.8125
4	1.0000	1.0000	0.9999	0.9997	0.9990	0.9976	0.9947	0.9898	0.9815	0.9688
$n=6, x=0$	0.7351	0.5314	0.3771	0.2621	0.1780	0.1176	0.0754	0.0467	0.0277	0.0156
1	0.9672	0.8857	0.7765	0.6554	0.5339	0.4202	0.3191	0.2333	0.1636	0.1094
2	0.9978	0.9842	0.9527	0.9011	0.8306	0.7443	0.6471	0.5443	0.4415	0.3438
3	0.9999	0.9987	0.9941	0.9830	0.9624	0.9295	0.8826	0.8208	0.7447	0.6563
4	1.0000	0.9999	0.9996	0.9984	0.9954	0.9891	0.9777	0.9590	0.9308	0.8906
5	1.0000	1.0000	1.0000	0.9999	0.9998	0.9993	0.9982	0.9959	0.9917	0.9844
$n=7, x=0$	0.6983	0.4783	0.3206	0.2097	0.1335	0.0824	0.0490	0.0280	0.0152	0.0078
1	0.9556	0.8503	0.7166	0.5767	0.4449	0.3294	0.2338	0.1586	0.1024	0.0625
2	0.9962	0.9743	0.9262	0.8520	0.7564	0.6471	0.5323	0.4199	0.3164	0.2266
3	0.9998	0.9973	0.9879	0.9667	0.9294	0.8740	0.8002	0.7102	0.6083	0.5000
4	1.0000	0.9998	0.9988	0.9953	0.9871	0.9712	0.9444	0.9037	0.8471	0.7734
5	1.0000	1.0000	0.9999	0.9996	0.9987	0.9962	0.9910	0.9812	0.9643	0.9375
6	1.0000	1.0000	1.0000	1.0000	0.9999	0.9998	0.9994	0.9984	0.9963	0.9922
$n=8, x=0$	0.6634	0.4305	0.2725	0.1678	0.1001	0.0576	0.0319	0.0168	0.0084	0.0039
1	0.9428	0.8131	0.6572	0.5033	0.3671	0.2553	0.1691	0.1064	0.0632	0.0352
2	0.9942	0.9619	0.8948	0.7969	0.6785	0.5518	0.4278	0.3154	0.2201	0.1445
3	0.9996	0.9950	0.9786	0.9437	0.8862	0.8059	0.7064	0.5941	0.4770	0.3633
4	1.0000	0.9996	0.9971	0.9896	0.9727	0.9420	0.8939	0.8263	0.7396	0.6367
5	1.0000	1.0000	0.9998	0.9988	0.9958	0.9887	0.9747	0.9502	0.9115	0.8555
6	1.0000	1.0000	1.0000	0.9999	0.9996	0.9987	0.9964	0.9915	0.9819	0.9648
7	1.0000	1.0000	1.0000	1.0000	1.0000	0.9999	0.9998	0.9993	0.9983	0.9961
$n=9, x=0$	0.6302	0.3874	0.2316	0.1342	0.0751	0.0404	0.0207	0.0101	0.0046	0.0020
1	0.9288	0.7748	0.5995	0.4362	0.3003	0.1960	0.1211	0.0705	0.0385	0.0195
2	0.9916	0.9470	0.8591	0.7382	0.6007	0.4628	0.3373	0.2318	0.1495	0.0898
3	0.9994	0.9917	0.9661	0.9144	0.8343	0.7297	0.6089	0.4826	0.3614	0.2539
4	1.0000	0.9991	0.9944	0.9804	0.9511	0.9012	0.8283	0.7334	0.6214	0.5000
5	1.0000	0.9999	0.9994	0.9969	0.9900	0.9747	0.9464	0.9006	0.8342	0.7461
6	1.0000	1.0000	1.0000	0.9997	0.9987	0.9957	0.9888	0.9750	0.9502	0.9102
7	1.0000	1.0000	1.0000	1.0000	0.9999	0.9996	0.9986	0.9962	0.9909	0.9805
8	1.0000	1.0000	1.0000	1.0000	1.0000	1.0000	0.9999	0.9997	0.9992	0.9980

$p=$	0.05	0.10	0.15	0.20	0.25	0.30	0.35	0.40	0.45	0.50
$n=10, x=0$	0.5987	0.3487	0.1969	0.1074	0.0563	0.0282	0.0135	0.0060	0.0025	0.0010
1	0.9139	0.7361	0.5443	0.3758	0.2440	0.1493	0.0860	0.0464	0.0233	0.0107
2	0.9885	0.9298	0.8202	0.6778	0.5256	0.3828	0.2616	0.1673	0.0996	0.0547
3	0.9990	0.9872	0.9500	0.8791	0.7759	0.6496	0.5138	0.3823	0.2660	0.1719
4	0.9999	0.9984	0.9901	0.9672	0.9219	0.8497	0.7515	0.6331	0.5044	0.3770
5	1.0000	0.9999	0.9986	0.9936	0.9803	0.9527	0.9051	0.8338	0.7384	0.6230
6	1.0000	1.0000	0.9999	0.9991	0.9965	0.9894	0.9740	0.9452	0.8980	0.8281
7	1.0000	1.0000	1.0000	0.9999	0.9996	0.9984	0.9952	0.9877	0.9726	0.9453
8	1.0000	1.0000	1.0000	1.0000	1.0000	0.9999	0.9995	0.9983	0.9955	0.9893
9	1.0000	1.0000	1.0000	1.0000	1.0000	1.0000	1.0000	0.9999	0.9997	0.9990
$n=12, x=0$	0.5404	0.2824	0.1422	0.0687	0.0317	0.0138	0.0057	0.0022	0.0008	0.0002
1	0.8816	0.6590	0.4435	0.2749	0.1584	0.0850	0.0424	0.0196	0.0083	0.0032
2	0.9804	0.8891	0.7358	0.5583	0.3907	0.2528	0.1513	0.0834	0.0421	0.0193
3	0.9978	0.9744	0.9078	0.7946	0.6488	0.4925	0.3467	0.2253	0.1345	0.0730
4	0.9998	0.9957	0.9761	0.9274	0.8424	0.7237	0.5833	0.4382	0.3044	0.1938
5	1.0000	0.9995	0.9954	0.9806	0.9456	0.8822	0.7873	0.6652	0.5269	0.3872
6	1.0000	0.9999	0.9993	0.9961	0.9857	0.9614	0.9154	0.8418	0.7393	0.6128
7	1.0000	1.0000	0.9999	0.9994	0.9972	0.9905	0.9745	0.9427	0.8883	0.8062
8	1.0000	1.0000	1.0000	0.9999	0.9996	0.9983	0.9944	0.9847	0.9644	0.9270
9	1.0000	1.0000	1.0000	1.0000	1.0000	0.9998	0.9992	0.9972	0.9921	0.9807
10	1.0000	1.0000	1.0000	1.0000	1.0000	1.0000	0.9999	0.9997	0.9989	0.9968
11	1.0000	1.0000	1.0000	1.0000	1.0000	1.0000	1.0000	1.0000	0.9999	0.9998
$n=15, x=0$	0.4633	0.2059	0.0874	0.0352	0.0134	0.0047	0.0016	0.0005	0.0001	0.0000
1	0.8290	0.5490	0.3186	0.1671	0.0802	0.0353	0.0142	0.0052	0.0017	0.0005
2	0.9638	0.8159	0.6042	0.3980	0.2361	0.1268	0.0617	0.0271	0.0107	0.0037
3	0.9945	0.9444	0.8227	0.6482	0.4613	0.2969	0.1727	0.0905	0.0424	0.0176
4	0.9994	0.9873	0.9383	0.8358	0.6865	0.5155	0.3519	0.2173	0.1204	0.0592
5	0.9999	0.9978	0.9832	0.9389	0.8516	0.7216	0.5643	0.4032	0.2608	0.1509
6	1.0000	0.9997	0.9964	0.9819	0.9434	0.8689	0.7548	0.6098	0.4522	0.3036
7	1.0000	1.0000	0.9994	0.9958	0.9827	0.9500	0.8868	0.7869	0.6535	0.5000
8	1.0000	1.0000	0.9999	0.9992	0.9958	0.9848	0.9578	0.9050	0.8182	0.6964
9	1.0000	1.0000	1.0000	0.9999	0.9992	0.9963	0.9876	0.9662	0.9231	0.8491
10	1.0000	1.0000	1.0000	1.0000	0.9999	0.9993	0.9972	0.9907	0.9745	0.9408
11	1.0000	1.0000	1.0000	1.0000	1.0000	0.9999	0.9995	0.9981	0.9937	0.9824
12	1.0000	1.0000	1.0000	1.0000	1.0000	1.0000	0.9999	0.9997	0.9989	0.9963
13	1.0000	1.0000	1.0000	1.0000	1.0000	1.0000	1.0000	1.0000	0.9999	0.9995
14	1.0000	1.0000	1.0000	1.0000	1.0000	1.0000	1.0000	1.0000	1.0000	1.0000

$p=$	0.05	0.10	0.15	0.20	0.25	0.30	0.35	0.40	0.45	0.50
$n=20, x=0$	0.3585	0.1216	0.0388	0.0115	0.0032	0.0008	0.0002	0.0000	0.0000	0.0000
1	0.7358	0.3917	0.1756	0.0692	0.0243	0.0076	0.0021	0.0005	0.0001	0.0000
2	0.9245	0.6769	0.4049	0.2061	0.0913	0.0355	0.0121	0.0036	0.0009	0.0002
3	0.9841	0.8670	0.6477	0.4114	0.2252	0.1071	0.0444	0.0160	0.0049	0.0013
4	0.9974	0.9568	0.8298	0.6296	0.4148	0.2375	0.1182	0.0510	0.0189	0.0059
5	0.9997	0.9887	0.9327	0.8042	0.6172	0.4164	0.2454	0.1256	0.0553	0.0207
6	1.0000	0.9976	0.9781	0.9133	0.7858	0.6080	0.4166	0.2500	0.1299	0.0577
7	1.0000	0.9996	0.9941	0.9679	0.8982	0.7723	0.6010	0.4159	0.2520	0.1316
8	1.0000	0.9999	0.9987	0.9900	0.9591	0.8867	0.7624	0.5956	0.4143	0.2517
9	1.0000	1.0000	0.9998	0.9974	0.9861	0.9520	0.8782	0.7553	0.5914	0.4119
10	1.0000	1.0000	1.0000	0.9994	0.9961	0.9829	0.9468	0.8725	0.7507	0.5881
11	1.0000	1.0000	1.0000	0.9999	0.9991	0.9949	0.9804	0.9435	0.8692	0.7483
12	1.0000	1.0000	1.0000	1.0000	0.9998	0.9987	0.9940	0.9790	0.9420	0.8684
13	1.0000	1.0000	1.0000	1.0000	1.0000	0.9997	0.9985	0.9935	0.9786	0.9423
14	1.0000	1.0000	1.0000	1.0000	1.0000	1.0000	0.9997	0.9984	0.9936	0.9793
15	1.0000	1.0000	1.0000	1.0000	1.0000	1.0000	1.0000	0.9997	0.9985	0.9941
16	1.0000	1.0000	1.0000	1.0000	1.0000	1.0000	1.0000	1.0000	0.9997	0.9987
17	1.0000	1.0000	1.0000	1.0000	1.0000	1.0000	1.0000	1.0000	1.0000	0.9998
18	1.0000	1.0000	1.0000	1.0000	1.0000	1.0000	1.0000	1.0000	1.0000	1.0000
$n=25, x=0$	0.2774	0.0718	0.0172	0.0038	0.0008	0.0001	0.0000	0.0000	0.0000	0.0000
1	0.6424	0.2712	0.0931	0.0274	0.0070	0.0016	0.0003	0.0001	0.0000	0.0000
2	0.8729	0.5371	0.2537	0.0982	0.0321	0.0090	0.0021	0.0004	0.0001	0.0000
3	0.9659	0.7636	0.4711	0.2340	0.0962	0.0332	0.0097	0.0024	0.0005	0.0001
4	0.9928	0.9020	0.6821	0.4207	0.2137	0.0905	0.0320	0.0095	0.0023	0.0005
5	0.9988	0.9666	0.8385	0.6167	0.3783	0.1935	0.0826	0.0294	0.0086	0.0020
6	0.9998	0.9905	0.9305	0.7800	0.5611	0.3407	0.1734	0.0736	0.0258	0.0073
7	1.0000	0.9977	0.9745	0.8909	0.7265	0.5118	0.3061	0.1536	0.0639	0.0216
8	1.0000	0.9995	0.9920	0.9532	0.8506	0.6769	0.4668	0.2735	0.1340	0.0539
9	1.0000	0.9999	0.9979	0.9827	0.9287	0.8106	0.6303	0.4246	0.2424	0.1148
10	1.0000	1.0000	0.9995	0.9944	0.9703	0.9022	0.7712	0.5858	0.3843	0.2122
11	1.0000	1.0000	0.9999	0.9985	0.9893	0.9558	0.8746	0.7323	0.5426	0.3450
12	1.0000	1.0000	1.0000	0.9996	0.9966	0.9825	0.9396	0.8462	0.6937	0.5000
13	1.0000	1.0000	1.0000	0.9999	0.9991	0.9940	0.9745	0.9222	0.8173	0.6550
14	1.0000	1.0000	1.0000	1.0000	0.9998	0.9982	0.9907	0.9656	0.9040	0.7878
15	1.0000	1.0000	1.0000	1.0000	1.0000	0.9995	0.9971	0.9868	0.9560	0.8852
16	1.0000	1.0000	1.0000	1.0000	1.0000	0.9999	0.9992	0.9957	0.9826	0.9461
17	1.0000	1.0000	1.0000	1.0000	1.0000	1.0000	0.9998	0.9988	0.9942	0.9784
18	1.0000	1.0000	1.0000	1.0000	1.0000	1.0000	1.0000	0.9997	0.9984	0.9927
19	1.0000	1.0000	1.0000	1.0000	1.0000	1.0000	1.0000	0.9999	0.9996	0.9980
20	1.0000	1.0000	1.0000	1.0000	1.0000	1.0000	1.0000	1.0000	0.9999	0.9995
21	1.0000	1.0000	1.0000	1.0000	1.0000	1.0000	1.0000	1.0000	1.0000	0.9999
22	1.0000	1.0000	1.0000	1.0000	1.0000	1.0000	1.0000	1.0000	1.0000	1.0000

$p=$	0.05	0.10	0.15	0.20	0.25	0.30	0.35	0.40	0.45	0.50
$n=30, x=0$	0.2146	0.0424	0.0076	0.0012	0.0002	0.0000	0.0000	0.0000	0.0000	0.0000
1	0.5535	0.1837	0.0480	0.0105	0.0020	0.0003	0.0000	0.0000	0.0000	0.0000
2	0.8122	0.4114	0.1514	0.0442	0.0106	0.0021	0.0003	0.0000	0.0000	0.0000
3	0.9392	0.6474	0.3217	0.1227	0.0374	0.0093	0.0019	0.0003	0.0000	0.0000
4	0.9844	0.8245	0.5245	0.2552	0.0979	0.0302	0.0075	0.0015	0.0002	0.0000
5	0.9967	0.9268	0.7106	0.4275	0.2026	0.0766	0.0233	0.0057	0.0011	0.0002
6	0.9994	0.9742	0.8474	0.6070	0.3481	0.1595	0.0586	0.0172	0.0040	0.0007
7	0.9999	0.9922	0.9302	0.7608	0.5143	0.2814	0.1238	0.0435	0.0121	0.0026
8	1.0000	0.9980	0.9722	0.8713	0.6736	0.4315	0.2247	0.0940	0.0312	0.0081
9	1.0000	0.9995	0.9903	0.9389	0.8034	0.5888	0.3575	0.1763	0.0694	0.0214
10	1.0000	0.9999	0.9971	0.9744	0.8943	0.7304	0.5078	0.2915	0.1350	0.0494
11	1.0000	1.0000	0.9992	0.9905	0.9493	0.8407	0.6548	0.4311	0.2327	0.1002
12	1.0000	1.0000	0.9998	0.9969	0.9784	0.9155	0.7802	0.5785	0.3592	0.1808
13	1.0000	1.0000	1.0000	0.9991	0.9918	0.9599	0.8737	0.7145	0.5025	0.2923
14	1.0000	1.0000	1.0000	0.9998	0.9973	0.9831	0.9348	0.8246	0.6448	0.4278
15	1.0000	1.0000	1.0000	0.9999	0.9992	0.9936	0.9699	0.9029	0.7691	0.5722
16	1.0000	1.0000	1.0000	1.0000	0.9998	0.9979	0.9876	0.9519	0.8644	0.7077
17	1.0000	1.0000	1.0000	1.0000	0.9999	0.9994	0.9955	0.9788	0.9286	0.8192
18	1.0000	1.0000	1.0000	1.0000	1.0000	0.9998	0.9986	0.9917	0.9666	0.8998
19	1.0000	1.0000	1.0000	1.0000	1.0000	1.0000	0.9996	0.9971	0.9862	0.9506
20	1.0000	1.0000	1.0000	1.0000	1.0000	1.0000	0.9999	0.9991	0.9950	0.9786
21	1.0000	1.0000	1.0000	1.0000	1.0000	1.0000	1.0000	0.9998	0.9984	0.9919
22	1.0000	1.0000	1.0000	1.0000	1.0000	1.0000	1.0000	1.0000	0.9996	0.9974
23	1.0000	1.0000	1.0000	1.0000	1.0000	1.0000	1.0000	1.0000	0.9999	0.9993
24	1.0000	1.0000	1.0000	1.0000	1.0000	1.0000	1.0000	1.0000	1.0000	0.9998
25	1.0000	1.0000	1.0000	1.0000	1.0000	1.0000	1.0000	1.0000	1.0000	1.0000

$p=$	0.05	0.10	0.15	0.20	0.25	0.30	0.35	0.40	0.45	0.50
$n=40, x=0$	0.1285	0.0148	0.0015	0.0001	0.0000	0.0000	0.0000	0.0000	0.0000	0.0000
1	0.3991	0.0805	0.0121	0.0015	0.0001	0.0000	0.0000	0.0000	0.0000	0.0000
2	0.6767	0.2228	0.0486	0.0079	0.0010	0.0001	0.0000	0.0000	0.0000	0.0000
3	0.8619	0.4231	0.1302	0.0285	0.0047	0.0006	0.0001	0.0000	0.0000	0.0000
4	0.9520	0.6290	0.2633	0.0759	0.0160	0.0026	0.0003	0.0000	0.0000	0.0000
5	0.9861	0.7937	0.4325	0.1613	0.0433	0.0086	0.0013	0.0001	0.0000	0.0000
6	0.9966	0.9005	0.6067	0.2859	0.0962	0.0238	0.0044	0.0006	0.0001	0.0000
7	0.9993	0.9581	0.7559	0.4371	0.1820	0.0553	0.0124	0.0021	0.0002	0.0000
8	0.9999	0.9845	0.8646	0.5931	0.2998	0.1110	0.0303	0.0061	0.0009	0.0001
9	1.0000	0.9949	0.9328	0.7318	0.4395	0.1959	0.0644	0.0156	0.0027	0.0003
10	1.0000	0.9985	0.9701	0.8392	0.5839	0.3087	0.1215	0.0352	0.0074	0.0011
11	1.0000	0.9996	0.9880	0.9125	0.7151	0.4406	0.2053	0.0709	0.0179	0.0032
12	1.0000	0.9999	0.9957	0.9568	0.8209	0.5772	0.3143	0.1285	0.0386	0.0083
13	1.0000	1.0000	0.9986	0.9806	0.8968	0.7032	0.4408	0.2112	0.0751	0.0192
14	1.0000	1.0000	0.9996	0.9921	0.9456	0.8074	0.5721	0.3174	0.1326	0.0403
15	1.0000	1.0000	0.9999	0.9971	0.9738	0.8849	0.6946	0.4402	0.2142	0.0769
16	1.0000	1.0000	1.0000	0.9990	0.9884	0.9367	0.7978	0.5681	0.3185	0.1341
17	1.0000	1.0000	1.0000	0.9997	0.9953	0.9680	0.8761	0.6885	0.4391	0.2148
18	1.0000	1.0000	1.0000	0.9999	0.9983	0.9852	0.9301	0.7911	0.5651	0.3179
19	1.0000	1.0000	1.0000	1.0000	0.9994	0.9937	0.9637	0.8702	0.6844	0.4373
20	1.0000	1.0000	1.0000	1.0000	0.9998	0.9976	0.9827	0.9256	0.7870	0.5627
21	1.0000	1.0000	1.0000	1.0000	1.0000	0.9991	0.9925	0.9608	0.8669	0.6821
22	1.0000	1.0000	1.0000	1.0000	1.0000	0.9997	0.9970	0.9811	0.9233	0.7852
23	1.0000	1.0000	1.0000	1.0000	1.0000	0.9999	0.9989	0.9917	0.9595	0.8659
24	1.0000	1.0000	1.0000	1.0000	1.0000	1.0000	0.9996	0.9966	0.9804	0.9231
25	1.0000	1.0000	1.0000	1.0000	1.0000	1.0000	0.9999	0.9988	0.9914	0.9597
26	1.0000	1.0000	1.0000	1.0000	1.0000	1.0000	1.0000	0.9996	0.9966	0.9808
27	1.0000	1.0000	1.0000	1.0000	1.0000	1.0000	1.0000	0.9999	0.9988	0.9917
28	1.0000	1.0000	1.0000	1.0000	1.0000	1.0000	1.0000	1.0000	0.9996	0.9968
29	1.0000	1.0000	1.0000	1.0000	1.0000	1.0000	1.0000	1.0000	0.9999	0.9989
30	1.0000	1.0000	1.0000	1.0000	1.0000	1.0000	1.0000	1.0000	1.0000	0.9997
31	1.0000	1.0000	1.0000	1.0000	1.0000	1.0000	1.0000	1.0000	1.0000	0.9999
32	1.0000	1.0000	1.0000	1.0000	1.0000	1.0000	1.0000	1.0000	1.0000	1.0000

$p=$	0.05	0.10	0.15	0.20	0.25	0.30	0.35	0.40	0.45	0.50
$n=50, x=0$	0.0769	0.0052	0.0003	0.0000	0.0000	0.0000	0.0000	0.0000	0.0000	0.0000
1	0.2794	0.0338	0.0029	0.0002	0.0000	0.0000	0.0000	0.0000	0.0000	0.0000
2	0.5405	0.1117	0.0142	0.0013	0.0001	0.0000	0.0000	0.0000	0.0000	0.0000
3	0.7604	0.2503	0.0460	0.0057	0.0005	0.0000	0.0000	0.0000	0.0000	0.0000
4	0.8964	0.4312	0.1121	0.0185	0.0021	0.0002	0.0000	0.0000	0.0000	0.0000
5	0.9622	0.6161	0.2194	0.0480	0.0070	0.0007	0.0001	0.0000	0.0000	0.0000
6	0.9882	0.7702	0.3613	0.1034	0.0194	0.0025	0.0002	0.0000	0.0000	0.0000
7	0.9968	0.8779	0.5188	0.1904	0.0453	0.0073	0.0008	0.0001	0.0000	0.0000
8	0.9992	0.9421	0.6681	0.3073	0.0916	0.0183	0.0025	0.0002	0.0000	0.0000
9	0.9998	0.9755	0.7911	0.4437	0.1637	0.0402	0.0067	0.0008	0.0001	0.0000
10	1.0000	0.9906	0.8801	0.5836	0.2622	0.0789	0.0160	0.0022	0.0002	0.0000
11	1.0000	0.9968	0.9372	0.7107	0.3816	0.1390	0.0342	0.0057	0.0006	0.0000
12	1.0000	0.9990	0.9699	0.8139	0.5110	0.2229	0.0661	0.0133	0.0018	0.0002
13	1.0000	0.9997	0.9868	0.8894	0.6370	0.3279	0.1163	0.0280	0.0045	0.0005
14	1.0000	0.9999	0.9947	0.9393	0.7481	0.4468	0.1878	0.0540	0.0104	0.0013
15	1.0000	1.0000	0.9981	0.9692	0.8369	0.5692	0.2801	0.0955	0.0220	0.0033
16	1.0000	1.0000	0.9993	0.9856	0.9017	0.6839	0.3889	0.1561	0.0427	0.0077
17	1.0000	1.0000	0.9998	0.9937	0.9449	0.7822	0.5060	0.2369	0.0765	0.0164
18	1.0000	1.0000	0.9999	0.9975	0.9713	0.8594	0.6216	0.3356	0.1273	0.0325
19	1.0000	1.0000	1.0000	0.9991	0.9861	0.9152	0.7264	0.4465	0.1974	0.0595
20	1.0000	1.0000	1.0000	0.9997	0.9937	0.9522	0.8139	0.5610	0.2862	0.1013
21	1.0000	1.0000	1.0000	0.9999	0.9974	0.9749	0.8813	0.6701	0.3900	0.1611
22	1.0000	1.0000	1.0000	1.0000	0.9990	0.9877	0.9290	0.7660	0.5019	0.2399
23	1.0000	1.0000	1.0000	1.0000	0.9996	0.9944	0.9604	0.8438	0.6134	0.3359
24	1.0000	1.0000	1.0000	1.0000	0.9999	0.9976	0.9793	0.9022	0.7160	0.4439
25	1.0000	1.0000	1.0000	1.0000	1.0000	0.9991	0.9900	0.9427	0.8034	0.5561
26	1.0000	1.0000	1.0000	1.0000	1.0000	0.9997	0.9955	0.9686	0.8721	0.6641
27	1.0000	1.0000	1.0000	1.0000	1.0000	0.9999	0.9981	0.9840	0.9220	0.7601
28	1.0000	1.0000	1.0000	1.0000	1.0000	1.0000	0.9993	0.9924	0.9556	0.8389
29	1.0000	1.0000	1.0000	1.0000	1.0000	1.0000	0.9997	0.9966	0.9765	0.8987
30	1.0000	1.0000	1.0000	1.0000	1.0000	1.0000	0.9999	0.9986	0.9884	0.9405
31	1.0000	1.0000	1.0000	1.0000	1.0000	1.0000	1.0000	0.9995	0.9947	0.9675
32	1.0000	1.0000	1.0000	1.0000	1.0000	1.0000	1.0000	0.9998	0.9978	0.9836
33	1.0000	1.0000	1.0000	1.0000	1.0000	1.0000	1.0000	0.9999	0.9991	0.9923
34	1.0000	1.0000	1.0000	1.0000	1.0000	1.0000	1.0000	1.0000	0.9997	0.9967
35	1.0000	1.0000	1.0000	1.0000	1.0000	1.0000	1.0000	1.0000	0.9999	0.9987
36	1.0000	1.0000	1.0000	1.0000	1.0000	1.0000	1.0000	1.0000	1.0000	0.9995
37	1.0000	1.0000	1.0000	1.0000	1.0000	1.0000	1.0000	1.0000	1.0000	0.9998
38	1.0000	1.0000	1.0000	1.0000	1.0000	1.0000	1.0000	1.0000	1.0000	1.0000

Percentage points of the Normal distribution

The values z in the table are those which a random variable $Z \sim N(0, 1)$ exceeds with probability p; that is, $P(Z > z) = 1 - \phi(z) = p$

p	z	p	z
0.5000	0.0000	0.0500	1.6449
0.4000	0.2533	0.0250	1.9600
0.3000	0.5244	0.0100	2.3263
0.2000	0.8416	0.0050	2.5758
0.1500	1.0364	0.0010	3.0902
0.1000	1.2816	0.0005	3.2905

Poisson cumulative distribution function

The tabulated value is $P(X \leq x)$, where X has a Poisson distribution with parameter λ

$\lambda=$	0.5	1.0	1.5	2.0	2.5	3.0	3.5	4.0	4.5	5.0
$x=0$	0.6065	0.3679	0.2231	0.1353	0.0821	0.0498	0.0302	0.0183	0.0111	0.0067
1	0.9098	0.7358	0.5578	0.4060	0.2873	0.1991	0.1359	0.0916	0.0611	0.0404
2	0.9856	0.9197	0.8088	0.6767	0.5438	0.4232	0.3208	0.2381	0.1736	0.1247
3	0.9982	0.9810	0.9344	0.8571	0.7576	0.6472	0.5366	0.4335	0.3423	0.2650
4	0.9998	0.9963	0.9814	0.9473	0.8912	0.8153	0.7254	0.6288	0.5321	0.4405
5	1.0000	0.9994	0.9955	0.9834	0.9580	0.9161	0.8576	0.7851	0.7029	0.6160
6	1.0000	0.9999	0.9991	0.9955	0.9858	0.9665	0.9347	0.8893	0.8311	0.7622
7	1.0000	1.0000	0.9998	0.9989	0.9958	0.9881	0.9733	0.9489	0.9134	0.8666
8	1.0000	1.0000	1.0000	0.9998	0.9989	0.9962	0.9901	0.9786	0.9597	0.9319
9	1.0000	1.0000	1.0000	1.0000	0.9997	0.9989	0.9967	0.9919	0.9829	0.9682
10	1.0000	1.0000	1.0000	1.0000	0.9999	0.9997	0.9990	0.9972	0.9933	0.9863
11	1.0000	1.0000	1.0000	1.0000	1.0000	0.9999	0.9997	0.9991	0.9976	0.9945
12	1.0000	1.0000	1.0000	1.0000	1.0000	1.0000	0.9999	0.9997	0.9992	0.9980
13	1.0000	1.0000	1.0000	1.0000	1.0000	1.0000	1.0000	0.9999	0.9997	0.9993
14	1.0000	1.0000	1.0000	1.0000	1.0000	1.0000	1.0000	1.0000	0.9999	0.9998
15	1.0000	1.0000	1.0000	1.0000	1.0000	1.0000	1.0000	1.0000	1.0000	0.9999
16	1.0000	1.0000	1.0000	1.0000	1.0000	1.0000	1.0000	1.0000	1.0000	1.0000
17	1.0000	1.0000	1.0000	1.0000	1.0000	1.0000	1.0000	1.0000	1.0000	1.0000
18	1.0000	1.0000	1.0000	1.0000	1.0000	1.0000	1.0000	1.0000	1.0000	1.0000
19	1.0000	1.0000	1.0000	1.0000	1.0000	1.0000	1.0000	1.0000	1.0000	1.0000

$\lambda=$	5.5	6.0	6.5	7.0	7.5	8.0	8.5	9.0	9.5	10.0
$x=0$	0.0041	0.0025	0.0015	0.0009	0.0006	0.0003	0.0002	0.0001	0.0001	0.0000
1	0.0266	0.0174	0.0113	0.0073	0.0047	0.0030	0.0019	0.0012	0.0008	0.0005
2	0.0884	0.0620	0.0430	0.0296	0.0203	0.0138	0.0093	0.0062	0.0042	0.0028
3	0.2017	0.1512	0.1118	0.0818	0.0591	0.0424	0.0301	0.0212	0.0149	0.0103
4	0.3575	0.2851	0.2237	0.1730	0.1321	0.0996	0.0744	0.0550	0.0403	0.0293
5	0.5289	0.4457	0.3690	0.3007	0.2414	0.1912	0.1496	0.1157	0.0885	0.0671
6	0.6860	0.6063	0.5265	0.4497	0.3782	0.3134	0.2562	0.2068	0.1649	0.1301
7	0.8095	0.7440	0.6728	0.5987	0.5246	0.4530	0.3856	0.3239	0.2687	0.2202
8	0.8944	0.8472	0.7916	0.7291	0.6620	0.5925	0.5231	0.4557	0.3918	0.3328
9	0.9462	0.9161	0.8774	0.8305	0.7764	0.7166	0.6530	0.5874	0.5218	0.4579
10	0.9747	0.9574	0.9332	0.9015	0.8622	0.8159	0.7634	0.7060	0.6453	0.5830
11	0.9890	0.9799	0.9661	0.9467	0.9208	0.8881	0.8487	0.8030	0.7520	0.6968
12	0.9955	0.9912	0.9840	0.9730	0.9573	0.9362	0.9091	0.8758	0.8364	0.7916
13	0.9983	0.9964	0.9929	0.9872	0.9784	0.9658	0.9486	0.9261	0.8981	0.8645
14	0.9994	0.9986	0.9970	0.9943	0.9897	0.9827	0.9726	0.9585	0.9400	0.9165
15	0.9998	0.9995	0.9988	0.9976	0.9954	0.9918	0.9862	0.9780	0.9665	0.9513
16	0.9999	0.9998	0.9996	0.9990	0.9980	0.9963	0.9934	0.9889	0.9823	0.9730
17	1.0000	0.9999	0.9998	0.9996	0.9992	0.9984	0.9970	0.9947	0.9911	0.9857
18	1.0000	1.0000	0.9999	0.9999	0.9997	0.9993	0.9987	0.9976	0.9957	0.9928
19	1.0000	1.0000	1.0000	1.0000	0.9999	0.9997	0.9995	0.9989	0.9980	0.9965
20	1.0000	1.0000	1.0000	1.0000	1.0000	0.9999	0.9998	0.9996	0.9991	0.9984
21	1.0000	1.0000	1.0000	1.0000	1.0000	1.0000	0.9999	0.9998	0.9996	0.9993
22	1.0000	1.0000	1.0000	1.0000	1.0000	1.0000	1.0000	0.9999	0.9999	0.9997

Percentage points of the χ^2 distribution

The values in the table are those which a random variable with the χ^2 distribution on v degrees of freedom exceeds with the probability shown.

v	0.995	0.990	0.975	0.950	0.900	0.100	0.050	0.025	0.010	0.005
1	0.000	0.000	0.001	0.004	0.016	2.705	3.841	5.024	6.635	7.879
2	0.010	0.020	0.051	0.103	0.211	4.605	5.991	7.378	9.210	10.597
3	0.072	0.115	0.216	0.352	0.584	6.251	7.815	9.348	11.345	12.838
4	0.207	0.297	0.484	0.711	1.064	7.779	9.488	11.143	13.277	14.860
5	0.412	0.554	0.831	1.145	1.610	9.236	11.070	12.832	15.086	16.750
6	0.676	0.872	1.237	1.635	2.204	10.645	12.592	14.449	16.812	18.548
7	0.989	1.239	1.690	2.167	2.833	12.017	14.067	16.013	18.475	20.278
8	1.344	1.646	2.180	2.733	3.490	13.362	15.507	17.535	20.090	21.955
9	1.735	2.088	2.700	3.325	4.168	14.684	16.919	19.023	21.666	23.589
10	2.156	2.558	3.247	3.940	4.865	15.987	18.307	20.483	23.209	25.188
11	2.603	3.053	3.816	4.575	5.580	17.275	19.675	21.920	24.725	26.757
12	3.074	3.571	4.404	5.226	6.304	18.549	21.026	23.337	26.217	28.300
13	3.565	4.107	5.009	5.892	7.042	19.812	22.362	24.736	27.688	29.819
14	4.075	4.660	5.629	6.571	7.790	21.064	23.685	26.119	29.141	31.319
15	4.601	5.229	6.262	7.261	8.547	22.307	24.996	27.488	30.578	32.801
16	5.142	5.812	6.908	7.962	9.312	23.542	26.296	28.845	32.000	34.267
17	5.697	6.408	7.564	8.672	10.085	24.769	27.587	30.191	33.409	35.718
18	6.265	7.015	8.231	9.390	10.865	25.989	28.869	31.526	34.805	37.156
19	6.844	7.633	8.907	10.117	11.651	27.204	30.144	32.852	36.191	38.582
20	7.434	8.260	9.591	10.851	12.443	28.412	31.410	34.170	37.566	39.997
21	8.034	8.897	10.283	11.591	13.240	29.615	32.671	35.479	38.932	41.401
22	8.643	9.542	10.982	12.338	14.042	30.813	33.924	36.781	40.289	42.796
23	9.260	10.196	11.689	13.091	14.848	32.007	35.172	38.076	41.638	44.181
24	9.886	10.856	12.401	13.848	15.659	33.196	36.415	39.364	42.980	45.558
25	10.520	11.524	13.120	14.611	16.473	34.382	37.652	40.646	44.314	46.928
26	11.160	12.198	13.844	15.379	17.292	35.563	38.885	41.923	45.642	48.290
27	11.808	12.879	14.573	16.151	18.114	36.741	40.113	43.194	46.963	49.645
28	12.461	13.565	15.308	16.928	18.939	37.916	41.337	44.461	48.278	50.993
29	13.121	14.256	16.047	17.708	19.768	39.088	42.557	45.722	49.588	52.336
30	13.787	14.953	16.791	18.493	20.599	40.256	43.773	46.979	50.892	53.672

Critical values for correlation coefficients

These tables concern tests of the hypothesis that a population correlation coefficient ρ is 0. The values in the tables are the minimum values which need to be reached by a sample correlation coefficient in order to be significant at the level shown, on a one-tailed test.

Product moment coefficient					Sample size, n	Spearman's coefficient		
Level						Level		
0.1	0.05	0.025	0.01	0.005		0.05	0.025	0.01
0.8000	0.9000	0.9500	0.9800	0.9900	4	1.0000	–	–
0.6870	0.8054	0.8783	0.9343	0.9587	5	0.9000	1.0000	1.0000
0.6084	0.7293	0.8114	0.8822	0.9172	6	0.8286	0.8857	0.9429
0.5509	0.6694	0.7545	0.8329	0.8745	7	0.7143	0.7857	0.8929
0.5067	0.6215	0.7067	0.7887	0.8343	8	0.6429	0.7381	0.8333
0.4716	0.5822	0.6664	0.7498	0.7977	9	0.6000	0.7000	0.7833
0.4428	0.5494	0.6319	0.7155	0.7646	10	0.5636	0.6485	0.7455
0.4187	0.5214	0.6021	0.6851	0.7348	11	0.5364	0.6182	0.7091
0.3981	0.4973	0.5760	0.6581	0.7079	12	0.5035	0.5874	0.6783
0.3802	0.4762	0.5529	0.6339	0.6835	13	0.4835	0.5604	0.6484
0.3646	0.4575	0.5324	0.6120	0.6614	14	0.4637	0.5385	0.6264
0.3507	0.4409	0.5140	0.5923	0.6411	15	0.4464	0.5214	0.6036
0.3383	0.4259	0.4973	0.5742	0.6226	16	0.4294	0.5029	0.5824
0.3271	0.4124	0.4821	0.5577	0.6055	17	0.4142	0.4877	0.5662
0.3170	0.4000	0.4683	0.5425	0.5897	18	0.4014	0.4716	0.5501
0.3077	0.3887	0.4555	0.5285	0.5751	19	0.3912	0.4596	0.5351
0.2992	0.3783	0.4438	0.5155	0.5614	20	0.3805	0.4466	0.5218
0.2914	0.3687	0.4329	0.5034	0.5487	21	0.3701	0.4364	0.5091
0.2841	0.3598	0.4227	0.4921	0.5368	22	0.3608	0.4252	0.4975
0.2774	0.3515	0.4133	0.4815	0.5256	23	0.3528	0.4160	0.4862
0.2711	0.3438	0.4044	0.4716	0.5151	24	0.3443	0.4070	0.4757
0.2653	0.3365	0.3961	0.4622	0.5052	25	0.3369	0.3977	0.4662
0.2598	0.3297	0.3882	0.4534	0.4958	26	0.3306	0.3901	0.4571
0.2546	0.3233	0.3809	0.4451	0.4869	27	0.3242	0.3828	0.4487
0.2497	0.3172	0.3739	0.4372	0.4785	28	0.3180	0.3755	0.4401
0.2451	0.3115	0.3673	0.4297	0.4705	29	0.3118	0.3685	0.4325
0.2407	0.3061	0.3610	0.4226	0.4629	30	0.3063	0.3624	0.4251
0.2070	0.2638	0.3120	0.3665	0.4026	40	0.2640	0.3128	0.3681
0.1843	0.2353	0.2787	0.3281	0.3610	50	0.2353	0.2791	0.3293
0.1678	0.2144	0.2542	0.2997	0.3301	60	0.2144	0.2545	0.3005
0.1550	0.1982	0.2352	0.2776	0.3060	70	0.1982	0.2354	0.2782
0.1448	0.1852	0.2199	0.2597	0.2864	80	0.1852	0.2201	0.2602
0.1364	0.1745	0.2072	0.2449	0.2702	90	0.1745	0.2074	0.2453
0.1292	0.1654	0.1966	0.2324	0.2565	100	0.1654	0.1967	0.2327

Percentage points of Student's *t* distribution

The values in the table are those which a random variable with Student's t distribution on ν degrees of freedom exceeds with the probability shown.

ν	0.1	0.05	0.025	0.01	0.005
1	3.078	6.314	12.706	31.821	63.657
2	1.886	2.920	4.303	6.965	9.925
3	1.638	2.353	3.182	4.541	5.841
4	1.533	2.132	2.776	3.747	4.604
5	1.476	2.015	2.571	3.365	4.032
6	1.440	1.943	2.447	3.143	3.707
7	1.415	1.895	2.365	2.998	3.499
8	1.397	1.860	2.306	2.896	3.355
9	1.383	1.833	2.262	2.821	3.250
10	1.372	1.812	2.228	2.764	3.169
11	1.363	1.796	2.201	2.718	3.106
12	1.356	1.782	2.179	2.681	3.055
13	1.350	1.771	2.160	2.650	3.012
14	1.345	1.761	2.145	2.624	2.977
15	1.341	1.753	2.131	2.602	2.947
16	1.337	1.746	2.120	2.583	2.921
17	1.333	1.740	2.110	2.567	2.898
18	1.330	1.734	2.101	2.552	2.878
19	1.328	1.729	2.093	2.539	2.861
20	1.325	1.725	2.086	2.528	2.845
21	1.323	1.721	2.080	2.518	2.831
22	1.321	1.717	2.074	2.508	2.819
23	1.319	1.714	2.069	2.500	2.807
24	1.318	1.711	2.064	2.492	2.797
25	1.316	1.708	2.060	2.485	2.787
26	1.315	1.706	2.056	2.479	2.779
27	1.314	1.703	2.052	2.473	2.771
28	1.313	1.701	2.048	2.467	2.763
29	1.311	1.699	2.045	2.462	2.756
30	1.310	1.697	2.042	2.457	2.750
32	1.309	1.694	2.037	2.449	2.738
34	1.307	1.691	2.032	2.441	2.728
36	1.306	1.688	2.028	2.435	2.719
38	1.304	1.686	2.024	2.429	2.712
40	1.303	1.684	2.021	2.423	2.704
45	1.301	1.679	2.014	2.412	2.690
50	1.299	1.676	2.009	2.403	2.678
55	1.297	1.673	2.004	2.396	2.668
60	1.296	1.671	2.000	2.390	2.660
70	1.294	1.667	1.994	2.381	2.648
80	1.292	1.664	1.990	2.374	2.639
90	1.291	1.662	1.987	2.369	2.632
100	1.290	1.660	1.984	2.364	2.626
110	1.289	1.659	1.982	2.361	2.621
120	1.289	1.658	1.980	2.358	2.617

Percentage points of the *F* distribution

The values in the table are those which a random variable with the *F* distribution on υ_1 and υ_2 degrees of freedom exceeds with probability 0.05 or 0.01

Probability	υ_2 \ υ_1	1	2	3	4	5	6	8	10	12	24	∞
0.05	1	161.40	199.50	215.70	224.60	230.20	234.00	238.90	241.90	243.90	249.10	254.30
	2	18.51	19.00	19.16	19.25	19.30	19.33	19.37	19.40	19.41	19.46	19.50
	3	10.13	9.55	9.28	9.12	9.01	8.94	8.85	8.79	8.74	8.64	8.53
	4	7.71	6.94	6.59	6.39	6.26	6.16	6.04	5.96	5.91	5.77	5.63
	5	6.61	5.79	5.41	5.19	5.05	4.95	4.82	4.74	4.68	4.53	4.37
	6	5.99	5.14	4.76	4.53	4.39	4.28	4.15	4.06	4.00	3.84	3.67
	7	5.59	4.74	4.35	4.12	3.97	3.87	3.73	3.64	3.57	3.41	3.23
	8	5.32	4.46	4.07	3.84	3.69	3.58	3.44	3.35	3.28	3.12	2.93
	9	5.12	4.26	3.86	3.63	3.48	3.37	3.23	3.14	3.07	2.90	2.71
	10	4.96	4.10	3.71	3.48	3.33	3.22	3.07	2.98	2.91	2.74	2.54
	11	4.84	3.98	3.59	3.36	3.20	3.09	2.95	2.85	2.79	2.61	2.40
	12	4.75	3.89	3.49	3.26	3.11	3.00	2.85	2.75	2.69	2.51	2.30
	14	4.60	3.74	3.34	3.11	2.96	2.85	2.70	2.60	2.53	2.35	2.13
	16	4.49	3.63	3.24	3.01	2.85	2.74	2.59	2.49	2.42	2.24	2.01
	18	4.41	3.55	3.16	2.93	2.77	2.66	2.51	2.41	2.34	2.15	1.92
	20	4.35	3.49	3.10	2.87	2.71	2.60	2.45	2.35	2.28	2.08	1.84
	25	4.24	3.39	2.99	2.76	2.60	2.49	2.34	2.24	2.16	1.96	1.71
	30	4.17	3.32	2.92	2.69	2.53	2.42	2.27	2.16	2.09	1.89	1.62
	40	4.08	3.23	2.84	2.61	2.45	2.34	2.18	2.08	2.00	1.79	1.51
	60	4.00	3.15	2.76	2.53	2.37	2.25	2.10	1.99	1.92	1.70	1.39
	120	3.92	3.07	2.68	2.45	2.29	2.18	2.02	1.91	1.83	1.61	1.25
	∞	3.84	3.00	2.60	2.37	2.21	2.10	1.94	1.83	1.75	1.52	1.00
0.01	1	4052.	5000.	5403.	5625.	5764.	5859.	5982.	6056.	6106.	6235.	6366.
	2	98.50	99.00	99.17	99.25	99.30	99.33	99.37	99.40	99.42	99.46	99.50
	3	34.12	30.82	29.46	28.71	28.24	27.91	27.49	27.23	27.05	26.60	26.13
	4	21.20	18.00	16.69	15.98	15.52	15.21	14.80	14.55	14.37	13.93	13.45
	5	16.26	13.27	12.06	11.39	10.97	10.67	10.29	10.05	9.89	9.47	9.02
	6	13.70	10.90	9.78	9.15	8.75	8.47	8.10	7.87	7.72	7.31	6.88
	7	12.20	9.55	8.45	7.85	7.46	7.19	6.84	6.62	6.47	6.07	5.65
	8	11.30	8.65	7.59	7.01	6.63	6.37	6.03	5.81	5.67	5.28	4.86
	9	10.60	8.02	6.99	6.42	6.06	5.80	5.47	5.26	5.11	4.73	4.31
	10	10.00	7.56	6.55	5.99	5.64	5.39	5.06	4.85	4.17	4.33	3.91
	11	9.65	7.21	6.22	5.67	5.32	5.07	4.74	4.54	4.40	4.02	3.60
	12	9.33	6.93	5.95	5.41	5.06	4.82	4.50	4.30	4.16	3.78	3.36
	14	8.86	6.51	5.56	5.04	4.70	4.46	4.14	3.94	3.80	3.43	3.00
	16	8.53	6.23	5.29	4.77	4.44	4.20	3.89	3.69	3.55	3.18	2.75
	18	8.29	6.01	5.09	4.58	4.25	4.01	3.71	3.51	3.37	3.00	2.57
	20	8.10	5.85	4.94	4.43	4.10	3.87	3.56	3.37	3.23	2.86	2.42
	25	7.77	5.57	4.68	4.18	3.86	3.63	3.32	3.13	2.99	2.62	2.17
	30	7.56	5.39	4.51	4.02	3.70	3.47	3.17	2.98	2.84	2.47	2.01
	40	7.31	5.18	4.31	3.83	3.51	3.29	2.99	2.80	2.66	2.29	1.80
	60	7.08	4.98	4.13	3.65	3.34	3.12	2.82	2.63	2.50	2.12	1.60
	120	6.85	4.79	3.95	3.48	3.17	2.96	2.66	2.47	2.34	1.95	1.38
	∞	6.63	4.61	3.78	3.32	3.02	2.80	2.51	2.32	2.18	1.79	1.00

If an *upper* percentage point of the *F* distribution on υ_1 and υ_2 degrees of freedom is *f*, then the corresponding *lower* percentage point of the *F* distribution on υ_2 and υ_1 degrees of freedom is $1/f$

Random numbers

32	35	58	30	97	44	79	40	31	25	68	78	56	69	06	76	04	51	79	46
70	84	67	80	76	68	41	73	78	75	70	33	49	56	13	99	62	09	69	37
38	95	09	80	43	33	26	60	55	45	96	11	97	08	29	10	48	08	31	64
72	56	02	04	84	14	14	27	50	38	91	84	01	82	62	34	47	71	42	97
87	90	15	49	08	47	89	14	21	19	93	20	23	30	71	43	06	42	10	13
94	22	42	79	72	90	25	74	17	63	75	16	09	66	06	23	92	23	87	72
20	07	52	57	92	08	15	21	90	75	21	80	74	74	23	10	49	38	94	79
75	91	49	57	59	50	79	29	72	04	07	70	78	92	99	78	90	13	52	26
28	70	08	90	90	95	39	32	58	04	11	83	54	28	72	78	82	68	97	93
46	05	87	83	06	61	95	37	94	94	20	06	17	48	47	72	90	27	61	51
36	88	61	53	11	72	92	25	85	35	03	43	75	15	27	31	69	95	73	43
41	85	76	03	91	16	92	62	81	86	26	17	41	46	28	96	12	80	87	88
11	85	08	80	91	92	27	45	49	33	99	17	84	68	29	92	89	80	32	36
19	52	12	48	64	70	29	06	81	50	78	17	04	75	45	31	97	53	52	46
20	08	69	83	22	27	12	13	98	02	67	18	27	32	35	13	76	10	11	30
94	43	14	15	77	84	89	73	19	45	22	09	78	36	13	99	15	04	77	98
70	36	60	28	91	50	86	07	45	59	53	19	68	21	62	67	46	06	67	56
95	06	58	20	75	73	55	91	22	21	88	71	34	89	86	18	99	91	23	22
58	75	25	03	00	28	32	10	35	30	20	25	53	95	48	98	78	89	99	93
25	38	28	08	69	61	69	75	09	87	64	90	98	09	43	10	41	27	77	45

Mathematical formulae – to learn
For AS and A Level Further Maths

You are expected to know the following mathematical formulae, and they will not be provided for you. Statistics formulae, for this course, are provided at the start. Formulae for the entire A Level course are then also provided for your reference.

Statistics

The mean of a set of data: $\bar{x} = \frac{\sum x}{n} = \frac{\sum fx}{\sum f}$

The standard Normal variable: $Z = \frac{X - \mu}{\sigma}$ where $X \sim N(\mu, \sigma^2)$

Pure Mathematics

Quadratic equations

$ax^2 + bx + c = 0$ has roots $\frac{-b \pm \sqrt{b^2 - 4ac}}{2a}$

Laws of indices

$a^x a^y \equiv a^{x+y}$

$a^x \div a^y \equiv a^{x-y}$

$(a^x)^y \equiv a^{xy}$

Laws of logarithms

$x = a^n \Leftrightarrow n = \log_a x$ for $a > 0$ and $x > 0$

$\log_a x + \log_a y \equiv \log_a xy$

$\log_a x - \log_a y \equiv \log_a \left(\frac{x}{y}\right)$

$k \log_a x \equiv \log_a (x)^k$

Coordinate geometry

A straight-line graph, gradient m passing through (x_1, y_1), has equation $y - y_1 = m(x - x_1)$

Straight lines with gradients m_1 and m_2 are perpendicular when $m_1 m_2 = -1$

Sequences

General term of an arithmetic progression

$$u_n = a + (n-1)d$$

General term of a geometric progression

$$u_n = ar^{n-1}$$

Trigonometry

In the triangle ABC:

Sine rule: $\dfrac{a}{\sin A} = \dfrac{b}{\sin B} = \dfrac{c}{\sin C}$

Cosine rule: $a^2 = b^2 + c^2 - 2bc\cos A$

Area: $\dfrac{1}{2}ab\sin C$

$$\cos^2 A + \sin^2 A \equiv 1$$

$$\sec^2 A \equiv 1 + \tan^2 A$$

$$\text{cosec}^2 A \equiv 1 + \cot^2 A$$

$$\sin 2A \equiv 2\sin A\cos A$$

$$\cos 2A \equiv \cos^2 A - \sin^2 A$$

$$\tan 2A \equiv \frac{2\tan A}{1-\tan^2 A}$$

Mensuration

Circumference and area of circle radius r and diameter d

$$C = 2\pi r = \pi d \qquad A = \pi r^2$$

Pythagoras' theorem:

In any right-angled triangle where a, b and c are the lengths of the sides and c is the hypotenuse, $c^2 = a^2 + b^2$

Area of a trapezium $= \dfrac{1}{2}(a+b)h$, where a and b are the lengths of the parallel sides and h is their perpendicular separation.

Volume of a prism = area of cross section × length

For a circle of radius r, where an angle at the centre of θ radians subtends an arc of length s and encloses an associated sector of area A

$$s = r\theta \qquad A = \frac{1}{2}r^2\theta$$

Complex numbers

For two complex numbers $z_1 = r_1 e^{i\theta_1}$ and $z_2 = r_2 e^{i\theta_2}$

$$z_1 z_2 = r_1 r_2\, e^{i(\theta_1 + \theta_2)}$$

$$\frac{z_1}{z_2} = \frac{r_1}{r_2} e^{i(\theta_1 - \theta_2)}$$

Loci in the Argand diagram:

$|z - a| = r$ is a circle radius r centred at a

$\arg(z - a) = \theta$ is a half line drawn from a at angle θ to a line parallel to the positive real axis.

Exponential Form: $e^{i\theta} = \cos\theta + i\sin\theta$

Matrices

For a 2 by 2 matrix $\begin{pmatrix} a & b \\ c & d \end{pmatrix}$ the determinant $\Delta = \begin{vmatrix} a & b \\ c & d \end{vmatrix} = ad - bc$

the inverse is $\frac{1}{\Delta}\begin{pmatrix} d & -b \\ -c & a \end{pmatrix}$

The transformation represented by matrix **AB** is the transformation represented by matrix **B** followed by the transformation represented by matrix **A**

For matrices **A**, **B**

$$(\mathbf{AB})^{-1} = \mathbf{B}^{-1}\mathbf{A}^{-1}$$

Algebra

$$\sum_{r=1}^{n} r = \frac{1}{2} n(n+1)$$

For $ax^2 + bx + c = 0$ with roots α and β

$$\alpha + \beta = -\frac{b}{a} \qquad \alpha\beta = \frac{c}{a}$$

For $ax^3 + bx^2 + cx + d = 0$ with roots α, β and γ

$$\sum\alpha = -\frac{b}{a} \qquad \sum\alpha\beta = \frac{c}{a} \qquad \alpha\beta\gamma = -\frac{d}{a}$$

Hyperbolic functions

$\cosh x \equiv \frac{1}{2}(e^{x} + e^{-x})$

$\sinh x \equiv \frac{1}{2}(e^{x} - e^{-x})$

$\tanh x \equiv \frac{\sinh x}{\cosh x}$

Calculus and differential equations

<u>Differentiation</u>

Function	Derivative
x^n	nx^{n-1}
$\sin kx$	$k\cos kx$
$\cos kx$	$-k\sin kx$
$\sinh kx$	$k\cosh kx$
$\cosh kx$	$k\sinh kx$
e^{kx}	ke^{kx}
$\ln x$	$\frac{1}{x}$
$f(x) + g(x)$	$f'(x) + g'(x)$
$f(x)g(x)$	$f'(x)g(x) + f(x)g'(x)$
$f(g(x))$	$f'(g(x))g'(x)$

Integration

Function	Integral
x^n	$\frac{1}{n+1}x^{n+1}+c,\ n\neq -1$
$\cos kx$	$\frac{1}{k}\sin kx+c$
$\sin kx$	$-\frac{1}{k}\cos kx+c$
$\cosh kx$	$\frac{1}{k}\sinh kx+c$
$\sinh kx$	$\frac{1}{k}\cosh kx+c$
e^{kx}	$\frac{1}{k}e^{kx}+c$
$\frac{1}{x}$	$\ln\|x\|+c,\ x\neq 0$
$f'(x)+g'(x)$	$f(x)+g(x)+c$
$f'(g(x))g'(x)$	$f(g(x))+c$

Area under a curve $= \int_a^b y\,dx\ (y \geq 0)$

Volumes of revolution about the x and y axes

$$V_x = \pi \int_a^b y^2\,dx \qquad V_y = \pi \int_c^d x^2\,dy$$

Simple harmonic motion

$$\ddot{x} = -\omega^2 x$$

Vectors

$|x\mathbf{i}+y\mathbf{j}+z\mathbf{k}| = \sqrt{(x^2+y^2+z^2)}$

Scalar product of two vectors $\mathbf{a} = \begin{pmatrix} a_1 \\ a_2 \\ a_3 \end{pmatrix}$ and $\mathbf{b} = \begin{pmatrix} b_1 \\ b_2 \\ b_3 \end{pmatrix}$ is

$$\begin{pmatrix} a_1 \\ a_2 \\ a_3 \end{pmatrix} \cdot \begin{pmatrix} b_1 \\ b_2 \\ b_3 \end{pmatrix} = a_1 b_1 + a_2 b_2 + a_3 b_3 = |\mathbf{a}|\,|\mathbf{b}| \cos\theta$$

where θ is the acute angle between the vectors **a** and **b**

The equation of the line through the point with position vector **a** parallel to vector **b** is

$$\mathbf{r} = \mathbf{a} + t\mathbf{b}$$

The equation of the plane containing the point with position vector **a** and perpendicular to vector **n** is

$$(\mathbf{r}-\mathbf{a}) \cdot \mathbf{n} = 0$$

Mechanics

Forces and equilibrium

Weight = mass $\times\, \mathbf{g}$

Friction: $\mathbf{F} \le \mu\mathbf{R}$

Newton's second law in the form: $\mathbf{F} = m\mathbf{a}$

Kinematics

For motion in a straight line with variable acceleration

$$\mathbf{v} = \frac{\mathrm{d}\mathbf{r}}{\mathrm{d}t} \qquad \mathbf{a} = \frac{\mathrm{d}\mathbf{v}}{\mathrm{d}t} = \frac{\mathrm{d}^2\mathbf{r}}{\mathrm{d}t^2}$$

$$\mathbf{r} = \int \mathbf{v}\,\mathrm{d}t \qquad \mathbf{v} = \int \mathbf{a}\,\mathrm{d}t$$

Momentum $= m\mathbf{v}$

Impulse $= m\mathbf{v} - m\mathbf{u}$

Kinetic energy $= \frac{1}{2} m\mathbf{v}^2$

Potential energy $= mgh$

The tension in an elastic string $= \frac{\lambda x}{l}$

The energy stored in an elastic string $= \frac{\lambda x^2}{2l}$

For simple harmonic motion

$$\ddot{x} = -\omega^2 x$$

$$x = a\cos\omega t \quad \text{or} \quad x = a\sin\omega t$$

$$v^2 = \omega^2(a^2 - x^2)$$

$$T = \frac{2\pi}{\omega}$$

Mathematical notation
For AS and A Level Further Maths

You should understand the following notation for AS and A Level Further Maths, without need for further explanation. Notation from all strands of mathematics has been included for your reference.

Probability and statistics

A, B, C, etc.	events
$A \cup B$	union of the events A and B
$A \cap B$	intersection of the events A and B
$\mathrm{P}(A)$	probability of the event A
A'	complement of the event A
$\mathrm{P}(A \mid B)$	probability of the event A conditional on the event B
X, Y, R, etc.	random variables
x, y, r, etc.	values of the random variables X, Y, R etc.
$x_1, x_2, \ldots$	observations
$f_1, f_2, \ldots$	frequencies with which the observations $x_1, x_2, \ldots$ occur
$\mathrm{p}(x)$, $\mathrm{P}(X = x)$	probability function of the discrete random variable X
$p_1, p_2, \ldots$	probabilities of the values $x_1, x_2, \ldots$ of the discrete random variable X
$\mathrm{E}(X)$	expectation of the random variable X
$\mathrm{Var}(X)$	variance of the random variable X
$\sim$	has the distribution
$\mathrm{B}(n, p)$	binomial distribution with parameters n and p, where n is the number of trials and p is the probability of success in a trial
q	$q = 1 - p$ for binomial distribution
$\mathrm{N}(\mu, \sigma^2)$	Normal distribution with mean μ and variance σ^2
$Z \sim \mathrm{N}(0,1)$	standard Normal distribution
ϕ	probability density function of the standardised Normal variable with distribution $\mathrm{N}(0, 1)$
Φ	corresponding cumulative distribution function
μ	population mean
σ^2	population variance
σ	population standard deviation
$\bar{x}$	sample mean
s^2	sample variance
s	sample standard deviation
H_0	null hypothesis
H_1	alternative hypothesis
r	product moment correlation coefficient for a sample
ρ	product moment correlation coefficient for a population
$\mathrm{Po}(\lambda)$	Poisson distribution with parameter λ
$\mathrm{Geo}(P)$	geometric distribution with parameter p
$\mathrm{G}_X(t)$	probability generating function of the random variable X
χ^2_v	chi squared distribution with v degrees of freedom
t_n	t distribution with n degrees of freedom
F_{v_1,v_2}	F distribution with v_1 and v_2 degree of freedom

Mathematical notation for AS and A Level Further Maths

Set notation

$\in$	is an element of
$\notin$	is not an element of
$\subseteq$	is a subset of
$\subset$	is a proper subset of
$\{x_1, x_2, \dots\}$	the set with elements $x_1, x_2, \dots$
$\{x: \dots\}$	the set of all x such that ...
$\mathrm{n}(A)$	the number of elements in set A
$\varnothing$	the empty set
ε	the universal set
A'	the complement of the set A
$\mathbb{N}$	the set of natural numbers, $\{1, 2, 3, \dots\}$
$\mathbb{Z}$	the set of integers, $\{0, \pm1, \pm2, \pm3, \dots\}$
$\mathbb{Z}^+$	the set of positive integers, $\{1, 2, 3, \dots\}$
$\mathbb{Z}_0^+$	the set of non-negative integers, $\{0, 1, 2, 3, \dots\}$
$\mathbb{R}$	the set of real numbers
$\mathbb{Q}$	the set of rational numbers, $\left\{\frac{p}{q} : p \in \mathbb{Z},\ q \in \mathbb{Z}^+\right\}$
$\cup$	union
$\cap$	intersection
(x, y)	the ordered pair x, y
$[a, b]$	the closed interval $\{x \in \mathbb{R} : a \le x \le b\}$
$[a, b)$	the interval $\{x \in \mathbb{R} : a \le x < b\}$
$(a, b]$	the interval $\{x \in \mathbb{R} : a < x \le b\}$
(a, b)	the open interval $\{x \in \mathbb{R} : a < x < b\}$
$\mathbb{C}$	the set of complex numbers

Miscellaneous symbols

$=$	is equal to
$\neq$	is not equal to
$\equiv$	is identical to or is congruent to
$\approx$	is approximately equal to
∞	infinity
$\propto$	is proportional to
$<$	is less than
$\leqslant, \le$	is less than or equal to; is not greater than
$>$	is greater than
$\geqslant, \ge$	is greater than or equal to; is not less than
$\therefore$	therefore
$\because$	because
$p \Rightarrow q$	p implies q (if p then q)
$p \Leftarrow q$	p is implied by q (if q then p)
$p \Leftrightarrow q$	p implies and is implied by q (p is equivalent to q)
a	first term for an arithmetic or geometric sequence
l	last term for an arithmetic sequence
d	common difference for an arithmetic sequence
r	common ratio for a geometric sequence
S_n	sum to n terms of a sequence
S_∞	sum to infinity of a sequence

Operations

$a+b$	a plus b
$a-b$	a minus b
$a\times b$, ab, $a\cdot b$	a multiplied by b
$a\div b$, $\frac{a}{b}$	a divided by b
$\sum_{i=1}^{n} a_i$	$a_1+a_2+\ldots+a_n$
$\prod_{i=1}^{n} a_i$	$a_1\times a_2\times\ldots\times a_n$
$\sqrt{a}$	the non-negative square root of a
$\|a\|$	the modulus of a
$n!$	n factorial: $n!=n\times(n-1)\times\ldots\times2\times1$, $n\in\mathbb{N}$; $0!=1$
$\binom{n}{r}$, ${}^nC_r$, ${}_nC_r$	the binomial coefficient $\frac{n!}{r!(n-r)!}$ for $n, r\in\mathbb{Z}_0^+$, $r\leq n$ or $\frac{n(n-1)\ldots(n-r+1)}{r!}$ for $n\in\mathbb{Q}$, $r\in\mathbb{Z}_0^+$

Functions

$\mathrm{f}(x)$	the value of the function f at x
$\lim_{x\to a}\mathrm{f}(x)$	the limit of $\mathrm{f}(x)$ as x tends to a
$\mathrm{f}: x\mapsto y$	the function f maps the element x to the element y
f^{-1}	the inverse function of the function f
gf	the composite function of f and g which is defined by $\mathrm{gf}(x)=\mathrm{g}(\mathrm{f}(x))$
Δx, δx	an increment of x
$\frac{\mathrm{d}y}{\mathrm{d}x}$	the derivative of y with respect to x
$\frac{\mathrm{d}^n y}{\mathrm{d}x^n}$	the nth derivative of y with respect to x
$\mathrm{f}'(x), \ldots, \mathrm{f}^{(n)}(x)$	the first, ..., nth derivatives of $\mathrm{f}(x)$ with respect to x
$\dot{x}, \ddot{x}, \ldots$	the first, second, ... derivatives of x with respect to t
$\int y\,\mathrm{d}x$	the indefinite integral of y with respect to x
$\int_a^b y\,\mathrm{d}x$	the definite integral of y with respect to x between the limits $x=a$ and $x=b$

Exponential and logarithmic functions

e	base of natural logarithms
e^x, $\exp x$	exponential function of x
$\log_a x$	logarithm to the base a of x
$\ln x$, $\log_\mathrm{e} x$	natural logarithm of x

Trigonometric functions

sin, cos, tan, cosec, sec, cot	the trigonometric functions
$\sin^{-1}$, $\cos^{-1}$, $\tan^{-1}$ arcsin, arccos, arctan	the inverse trigonometric functions
°	degrees
rad	radians
cosec^{-1}, $\sec^{-1}$, $\cot^{-1}$ arccosec, arcsec, arccot	the inverse trigonometric functions
sinh, cosh, tanh cosech, sech, coth	the hyperbolic functions
$\sinh^{-1}$, $\cosh^{-1}$, $\tanh^{-1}$ cosech^{-1}, sech^{-1}, $\coth^{-1}$ arsinh, arcosh, artanh arcosech, arsech, arcoth	the inverse hyperbolic functions

Complex numbers

i, j	square root of -1
$x + iy$	complex number with real part x and imaginary part y
$r(\cos\theta + i\sin\theta)$	modulus argument form of a complex number with modulus r and argument θ
z	a complex number $z = x + iy = r(\cos\theta + i\sin\theta)$
$\text{Re}(z)$	the real part of z, $\text{Re}(z) = x$
$\text{Im}(z)$	the imaginary part of z, $\text{Im}(z) = y$
$\lvert z\rvert$	the modulus of z, $\lvert z\rvert = \sqrt{x^2 + y^2}$
$\arg(z)$	the argument of z, $\arg(z) = \theta$, $-\pi < \theta < \pi$
z^*	the complex conjugate of z, $x - iy$

Matrices

$\mathbf{M}$	a matrix $\mathbf{M}$
$\mathbf{0}$	zero matrix
$\mathbf{I}$	identity matrix
$\mathbf{M}^{-1}$	the inverse of the matrix $\mathbf{M}$
$\mathbf{M}^{\text{T}}$	the transpose of the matrix $\mathbf{M}$
$\det \mathbf{M}$ or $\lvert\mathbf{M}\rvert$	the determinant of the square matrix $\mathbf{M}$
$\mathbf{Mr}$	Image of the column vector $\mathbf{r}$ under the transformation associated with the matrix $\mathbf{M}$

Vectors

$\mathbf{a}$, $\underline{a}$, $\underset{\sim}{a}$	the vector $\mathbf{a}$, $\underline{a}$, $\underset{\sim}{a}$
$\overrightarrow{AB}$	the vector represented in magnitude and direction by the directed line segment AB
$\hat{\mathbf{a}}$	a unit vector in the direction of $\mathbf{a}$
$\mathbf{i}, \mathbf{j}, \mathbf{k}$	unit vectors in the directions of the Cartesian coordinate axes
$\lvert\mathbf{a}\rvert$, a	the magnitude of $\mathbf{a}$
$\lvert\overrightarrow{AB}\rvert$, AB	the magnitude of $\overrightarrow{AB}$
$\begin{pmatrix} a \\ b \end{pmatrix}$, $a\mathbf{i} + b\mathbf{j}$	column vector and corresponding unit vector notation
$\mathbf{r}$	position vector
$\mathbf{s}$	displacement vector
$\mathbf{v}$	velocity vector
$\mathbf{a}$	acceleration vector
$\mathbf{a} \cdot \mathbf{b}$	the scalar product of $\mathbf{a}$ and $\mathbf{b}$
$\mathbf{a} \times \mathbf{b}$	the vector product of $\mathbf{a}$ and $\mathbf{b}$
$\mathbf{a}.\mathbf{b} \times \mathbf{c}$	the scalar triple product of $\mathbf{a}$, $\mathbf{b}$ and $\mathbf{c}$

Differential equations

ω	angular speed

Mechanics

kg	kilograms
m	metres
km	kilometres
m/s, m s^{-1}	metres per second (velocity)
m/s^2, m s^{-2}	metres per second squared (acceleration)
F	Force or resultant force
N	Newton
N m	Newton metre (moment of a force)
t	time
s	displacement
u	initial velocity
v	velocity or final velocity
a	acceleration
g	acceleration due to gravity
μ	coefficient of friction

Answers

Exercise 1.1A

1 a $E(X) = \frac{1}{20}$

$Var(X) = \frac{5369}{36} = 149$ (3 sf)

b $E(Y) = \frac{1}{6}$

$Var(Y) = \frac{41}{36}$

2 a $E(X) = \frac{5}{2}$

$Var(X) = \frac{5}{4}$

b $E(X) = \frac{5}{3}$

$Var(X) = \frac{2}{9}$

c $E(X) = \frac{8}{3}$

$Var(X) = \frac{2}{3}$

3 $E(X) = 1\times\frac{1}{10} + 2\times\frac{1}{10} + \ldots + 10\times\frac{1}{10} = \frac{11}{2}$

$Var(X) = 1^2\times\frac{1}{10} + 2^2\times\frac{1}{10} + \ldots + 10^2\times\frac{1}{10} - \left(\frac{11}{2}\right)^2 = \frac{33}{4}$

4 6

5 $E(\text{score}) = \frac{7}{3}$

$Var(\text{score}) = \frac{11}{9}$

6 $E(X) = 1\times\frac{1}{9} + 2\times\frac{1}{9} + \ldots + 9\times\frac{1}{9} = \frac{45}{9} = 5$

$Var(X) = \frac{20}{3}$

7 $\frac{91}{6}$

8 $\frac{63}{13} = 4.85$ (3 sf)

9 $\frac{49}{120} = 0.408$ (3 sf)

10 $E(X) = 2.2$

$Var(X) = 1.36$

11 a $\frac{1}{3}$

b If the experiment is performed a large number of times, approximately $\frac{1}{3}$ of the results will be prime. The proportion will get closer to $\frac{1}{3}$ as the number of trials increases.

12 a

x	1	2	3	4
P(X = x)	$\frac{1}{2}$	$\frac{1}{4}$	$\frac{1}{8}$	$\frac{1}{8}$

b $\frac{15}{8}$

Exercise 1.1B

1 $a = \frac{1}{5}$, $b = \frac{2}{15}$

2 a Mean = 0

Standard deviation = 2

b $\frac{5}{7}$

3 a $X \sim B(4, 0.49)$

b $E(X) = 1.96$ (3 sf)

$Var(X) = 0.9996$ (4 sf)

4 $n = 18$, $p = 0.45$

5 a $\frac{671}{1296} = 0.518$ (3 sf)

b 48p

6 $\frac{4}{3}$

7 a $E(X) = \frac{1}{5n}\sum_{x=1}^{5n} x = \frac{5n(5n+1)}{2\times 5n} = \frac{1+5n}{2}$

b $\frac{25n^2 - 1}{12}$

8 a 2.37 (3 sf)

b 2.84

Exercise 1.2A

1 0.01832, 0.07326, 0.1465, 0.1954, 0.5665

2 0.8009

3 0.389 (3 sf)

4 a

x	0	1	2	3	4	>4
P(X = x)	0.333	0.366	0.201	0.074	0.020	0.006

b

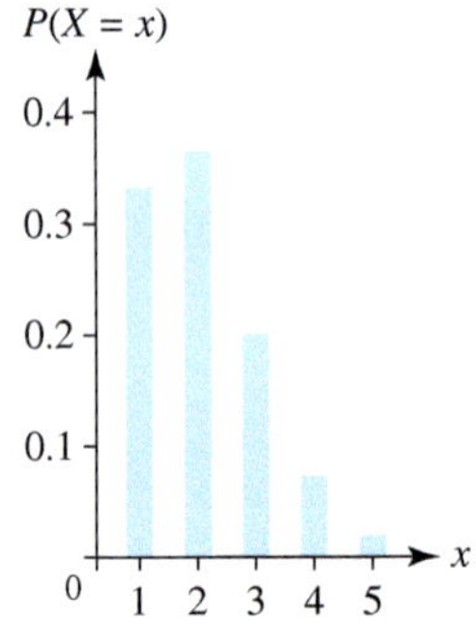

5 0.00277 (3 sf)

6 0.347 (3 sf)

7 0.879 (3 sf)

Assume random calls in time and therefore a Poisson distribution. Assume independence from one day to the next.

8 0.762 (3 sf)

9 0.912 (3 sf)

Distribution of hazelnuts is random. Number in any bar independent of any other bar.

10 0.00111 (3 sf)

11 0.195 (3 sf)

Exercise 1.2B

1 0.347 (3 sf)
2 0.195 (3 sf)
3 0.841 (3 sf)
4 0.142 (3 sf)
5 **a** n, 60, is large and p, 0.02, is small.
$X \sim \text{Po}(1.2)$
b **i** 0.2169
ii 0.1205
iii 0.1202
6 **a** Mean = 0.75
Variance = 0.7425
b Mean = 0.75
Variance = 0.75
7 **a** 0.122 (3 sf)
b 0.931 (3 sf)
8 **a** 0.148 (3 sf)
b 0.453 (3 sf)
c 0.0492 (3 sf)
9 **a** 0.545 (3 sf)
b 0.696 (3 sf)
c 0.774
10 0.00674 (3 sf)
Drops arrive randomly and evenly across the plate.
11 0.145 (3 sf)
12 **a** 0.265 (3sf)
b 0.0137 (3 sf)

Exercise 1.3A

1 **a** 0.25
b $\frac{63}{64} = 0.984$ (3 sf)
2 **a** 0.09
b 0.0009
c 0.99
d 0.01
3 **a**

x	1	2	3	4	5	6
$P(X = x)$	0.25	0.188	0.141	0.105	0.079	0.059

b
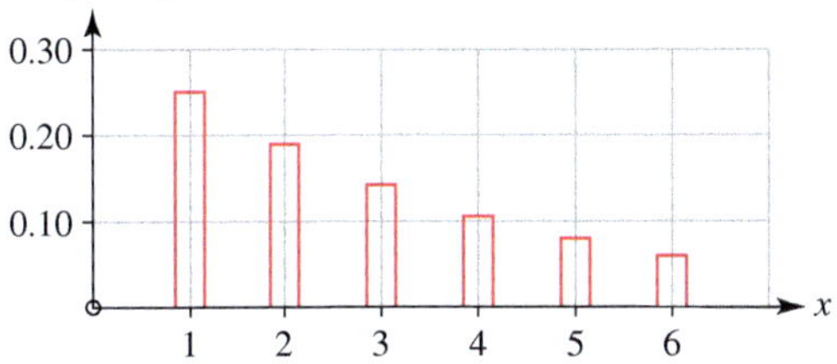

4 **a** $P(X = x) = 0.6^{x-1}0.4$
b $F(a) = P(X \le a) = \sum_{x=1}^{a}(1-p)^{x-1}p = \sum_{x=1}^{a}0.6^{x-1}0.4$
$= 0.4 + 0.6 \times 0.4 + 0.6^2 \times 0.4 + \ldots + 0.6^{a-1} \times 0.4$
$= \frac{0.4(1-0.6^a)}{1-0.6} = 1 - 0.6^a$
c $E(X) = 2.5$
$\text{Var}(X) = 3.75$
5 **a** $E(X) = 4$
Mode = 1
Standard deviation = 3.46 (3 sf)
b $\frac{\sqrt{3}}{2}$ or 0.866 (3 sf)

6 **a** $P(X < a) = \sum_{x=1}^{a-1}(1-p)^{x-1}p = p + qp + q^2p + \ldots + q^{a-2}p$
where $q = 1 - p$
$= \frac{p(1-q^{a-1})}{1-q} = 1 - q^{a-1}$
b $P(X \ge a) = 1 - P(X < a) = 1 - (1 - q^{a-1}) = q^{a-1}$
7 **a** $P(X = x) = (1-p)^{x-1}p$; $x = 1, 2, 3, \ldots$
$F(a) = P(X \le a) = \sum_{x=1}^{a}(1-p)^{x-1}p$
$= p + qp + q^2p + \ldots + q^{a-1}p$ where $q = 1 - p$
$= \frac{p(1-q^a)}{1-q} = 1 - q^a$
b 2
8 $(1-p)^{x-1}p$; $x = 1, 2, 3, \ldots$ $P(S) = p + (1-p)p + (1-p)^2p + \ldots$
$= \frac{p}{1-(1-p)} = 1$

Exercise 1.3B

1 **a** $X \sim \text{Geometric}\left(\frac{1}{6}\right)$
b **i** $\frac{125}{1296} = 0.0965$ (3 sf)
ii $\frac{25}{36} = 0.694$ (3 sf)
c $E(X) = 6$
$\text{Var}(X) = 30$
2 **a** 6
b $\frac{125}{1296} = 0.0965$ (3 sf)
3 Mean = 2
Variance = 2
4 **a** Let X be the number of prizes obtained from five packets.
$X \sim B(5, 0.12)$
0.360 (3 sf)
b Let Y be the number of packets until the first prize.
$Y \sim \text{Geometric}(0.12)$
0.0720 (3 sf)
5 **a** Let X be the number of people approached.
$X \sim \text{Geometric}(p)$ where p is the probability that a randomly chosen person will respond.
$\frac{1}{8}$
b $\frac{49}{64} = 0.766$ (3 sf)
6 **a** $P(Y = y) = (1-p)^y p$, $y = 0, 1, 2, 3, \ldots$
b $\frac{1-p}{p}$
c $E(Y) = E(X - 1) = E(X) - 1$
$= \frac{1}{p} - 1 = \frac{1-p}{p}$
7 **a** The probability p of a goal in a given minute is the same for every minute of the game. The number of goals in any given minute is independent of the outcome in any other minute.
b $P(G = g) = (1-p)^g p$; $g = 0, 1, 2, \ldots$
c 0.0274 (3 sf)

8 a $\text{lhs} = \frac{P(X > n+m \text{ and } X > m)}{P(X > m)} = \frac{P(X > n+m)}{P(X > m)} = \frac{q^{n+m}}{q^m} = q^n$,

$\text{rhs} = q^n$, lhs = rhs

b Result in **a** shows that the probability of getting an additional n failures, having already observed m failures, is the same as the probability of observing n failures at the start of the sequence; that is, the distribution 'forgets' what has occurred.

Exercise 1.4A

1 a 0.0953 (3 sf)

b 0.806 (3 sf)

2 a

x	3	4	5	6	7	8
$P(X=x)$	0.512	0.307	0.123	0.041	0.012	0.003

b

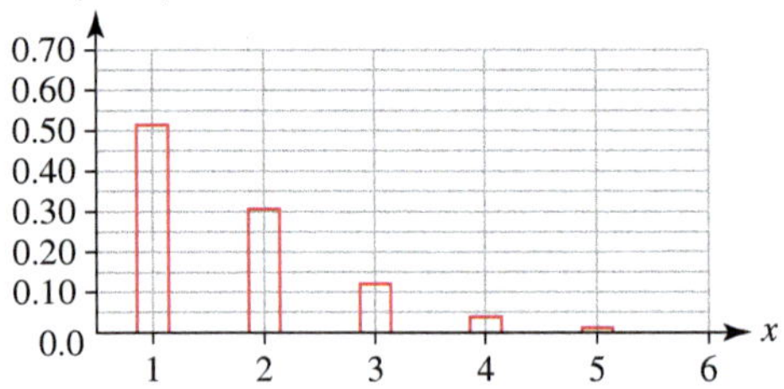

3 a $P(X=x) = \binom{x-1}{4} 0.8^5 0.2^{x-5}$; $x = 5, 6\ldots$

b 6.25; 1.56 (3 sf)

c i $P(X=6) = \frac{1024}{3125} = 0.328$ (3sf)

ii 1

iii $\frac{2048}{3125} = 0.655$ (3 sf)

4 a $P(X=x) = \binom{x-1}{5} 0.046656(0.4)^{x-6}$; $x = 6, 7, 8, \ldots$;

10; $\frac{20}{3} = 6.67$ (3 sf)

b 10

5 a $P(X=x) = \binom{x-1}{r-1} p^r (1-p)^{x-r}$; $x = r, (r+1)\ldots$

b $\text{Mean} = \frac{r}{p}$

$\text{Var} = \frac{r(1-p)}{p^2}$

c 0.633 (3 sf)

6 a $P(X=x) = \binom{x-1}{r-1} p^r (1-p)^{x-r}$; $x = r, (r+1)\ldots$;

$\frac{r}{p}, \frac{r(1-p)}{p^2}$

b If $r = 1$, X is the number of trials up to and including the 1st success, a geometric random variable.

c $P(X=x) = \binom{x-1}{0} p(1-p)^{x-1} = p(1-p)^{x-1}$; $\frac{r}{p} = \frac{1}{p}$;

$\frac{r(1-p)}{p^2} = \frac{(1-p)}{p^2}$

7 $P(X=r) = \binom{r-1}{r-1} p^r (1-p)^0 = p^r$;

$P(X=r+1) = \binom{r}{r-1} p^r (1-p)^1 = rp^r(1-p)$

$P(X=r) = P(X=r+1) \to p^r = rp^r(1-p) \Rightarrow r = \frac{1}{(1-p)}$

Exercise 1.4B

1 a Negative Binomial, parameters $r = 3, p = \frac{1}{6}$

$P(X=x) = \binom{x-1}{2}\left(\frac{1}{6}\right)^3\left(\frac{5}{6}\right)^{x-3}$; $x = 3, 4\ldots$

b 18; 90

c 0.0409 (3 sf)

2 a $X \sim$ Negative Binomial(6, 0.2)

b 30

c £18

On average, cheaper than buying tickets outright (£30)

3 a $X \sim$ Negative Binomial(5, 0.08)

b 62.5; 718.75

c 0.774 (3 sf)

4 a i 0.0477 (3 sf)

ii 0.150 (3 sf)

b $P(X=x) = \binom{x-1}{4} 0.56^5 0.44^{x-5}$

8.93 (3 sf); 7.02 (3 sf)

5 a The number of trials up to and including the $(r+s)$th success where the probability of success in any trial is p.

b Negative binomial parameters $r+s, p$

c $\frac{r+s}{p}$; $\frac{(r+s)(1-p)}{p^2}$

6 $P \sim$ Negative Binomial$\left(5, \frac{1}{2}\right)$, $Q \sim$ Negative Binomial$\left(7, \frac{1}{2}\right)$

and $R \sim$ Negative Binomial$\left(12, \frac{1}{2}\right)$

P(5th head on 8th toss|12th head on 24th toss)

$$= \frac{\text{P(5th head on 8th toss and 12th head on 24th toss)}}{\text{P(12th head on 24th toss)}}$$

$$= \frac{P(P=8) \times P(Q=16)}{P(R=24)}$$

$$= \frac{\binom{7}{4}\left(\frac{1}{2}\right)^5\left(\frac{1}{2}\right)^3 \times \binom{15}{6}\left(\frac{1}{2}\right)^7\left(\frac{1}{2}\right)^9}{\binom{23}{11}\left(\frac{1}{2}\right)^{12}\left(\frac{1}{2}\right)^{12}}$$

$$= \frac{\binom{7}{4}\binom{15}{6}}{\binom{23}{11}}$$

Exercise 1.5A

1 a $E(X) = 3$

$\text{Var}(X) = 2$

$\frac{2}{5}$

b 0.171 (3 sf)

2 a $\frac{2}{k}+\frac{3}{k}=1 \to k=5$

$E(X)=\frac{13}{5}$; $Var(X)=\frac{6}{25}$

b 0.886 (3 sf)

3 a $\frac{1}{a}+\frac{4}{a}+\frac{9}{a}+\frac{16}{a}=1 \to a=30$

b $\bar{X} \sim N(3.3333, 0.0164)$

0.149 (3 sf)

4 a $E(X)=\frac{13}{3}$

$Var(X)=\frac{20}{9}$

b 0.103 (3 sf)

5 a $E(X)=\frac{5}{2}$; $Var(X)=\frac{5}{4}$

b

a	1	$1\frac{1}{2}$	2	$2\frac{1}{2}$	3	$3\frac{1}{2}$	4
$P(\bar{X}=a)$	$\frac{1}{16}$	$\frac{2}{16}$	$\frac{3}{16}$	$\frac{4}{16}$	$\frac{3}{16}$	$\frac{2}{16}$	$\frac{1}{16}$

$\frac{3}{8}$

c 0.376 (3 sf)

d CLT can give a very good approximation for small sample sizes, especially if underlying distribution is nearly symmetric.

Exercise 1.5B

1 a $\bar{X} \sim N(3.6, 0.048)$

b 0.966 (3 sf)

2 a $X \sim Po(60)$

b $\bar{X} \sim N\left(60, \frac{10}{7}\right)$

c 0.953 (3 sf)

3 a $E(X_i)=p$; $Var(X_i)=p(1-p)$

b $\bar{X} \sim N\left(p, \frac{p(1-p)}{n}\right)$

c $X_T \sim B(n, p)$

d $X_T \sim N(np, np(1-p))$

4 a Geometric$\left(p=\frac{1}{13}\right)$.

$P(X=x)=\left(\frac{12}{13}\right)^{x-1}\left(\frac{1}{13}\right)$; $x=1, 2, 3, \ldots$

b $E(X)=13$

$Var(X)=156$

c 0.587 (3 sf)

5 a $X \sim Po\left(\frac{1}{4}\right)$

b i 0.1947 (4 dp)

ii 0.0265 (4 dp)

c $Y \sim$ Negative Binomial (3, 0.1947)

$E(Y)=15.4083$(4 dp); $Var(Y)=63.7305$ (4 dp)

Because $P(X>1)$ is small, 5s intervals can be treated as Bernoulli trials.

d $\bar{Y} \sim N(15.4083, 2.1244)$ (4 dp)

0.773 (3 sf)

Exercise 1.6A

1 $G_X(t)=0.3t+0.4t^2+0.3t^3$

2 $G_X(t)=\frac{5}{14}t+\frac{5}{14}t^2+\frac{2}{7}t^4$

3 a $\frac{1}{30}$

b $G_S(t)=\frac{1}{30}t+\frac{2}{15}t^2+\frac{3}{10}t^3+\frac{8}{15}t^4$

$E(S)=\frac{10}{3}$

$Var(S)=\frac{31}{45}$

4 a $G_R(t)=\frac{2}{k}t^2+\frac{3}{k}t^3+\frac{4}{k}t^4+\frac{5}{k}t^5$

b 14

$E(R)=\frac{27}{7}$; $Var(R)=\frac{55}{49}$

5 a $\frac{1}{4}$

b $P(X=x)=\left(\frac{3}{4}\right)^{x-1}\left(\frac{1}{4}\right)$

$\frac{1}{4}$; $\frac{3}{16}$; $\frac{9}{64}$

c $\mu=4$

$\sigma^2=12$

6 $P(X=x)=\frac{e^{-\lambda}\lambda^x}{x!}$; $x=0, 1, 2, \ldots$

$$G_X(t)=E\left[t^X\right]=\sum_{x=0}^{\infty}\frac{e^{-\lambda}\lambda^x}{x!}t^x=e^{-\lambda}\sum_{x=0}^{\infty}\frac{(\lambda t)^x}{x!}$$

$$=e^{-\lambda}\left(1+\lambda t+\frac{(\lambda t)^2}{2!}+\ldots\right)$$

$$=e^{-\lambda}e^{\lambda t}=e^{\lambda(t-1)}$$

$G'_X(t)=\lambda e^{\lambda(t-1)}$; $G''_X(t)=\lambda^2 e^{\lambda(t-1)}$

$\mu_X=G'_X(1)=\lambda$ and $\sigma^2=G''_X(1)+\mu-\mu^2=\lambda^2+\lambda-\lambda^2=\lambda$

7 a $G_X(t)=E\left[t^X\right]=\sum_{i=1}^{n}p_i t^{x_i}$

b $G_X(1)=\sum_{i=1}^{n}p_i=1$

c $G'_X(t)=\frac{d}{dt}G_X(t)=\frac{d}{dt}\sum_{i=1}^{n}p_i t^{x_i}=\sum_{i=1}^{n}p_i x_i t^{x_i-1}$

$G'_X(1)=\sum_{i=1}^{n}p_i x_i$

$G''_X(t)=\frac{d}{dt}G'_X(t)=\sum_{i=1}^{n}p_i x_i(x_i-1)t^{x_i-2}$

$G''_X(1)=\sum_{i=1}^{n}p_i x_i(x_i-1)$

d $\mu = \sum_{i=1}^{n} p_i x_i = G'_X(1)$

$$G''_X(1) = \sum_{i=1}^{n}(p_i x_i^2 - p_i x_i)$$
$$= \sum_{i=1}^{n} p_i x_i^2 - \sum_{i=1}^{n} p_i x_i$$
$$= \sum_{i=1}^{n} p_i x_i^2 - \mu^2 + \mu^2 - \mu$$
$$= \sigma^2 + \mu^2 - \mu \text{ because } \sigma^2 = \sum_{i=1}^{n} p_i x_i^2 - \mu^2$$
$$\sigma^2 = G''_X(1) + \mu - \mu^2$$

8 $G_Y(t) = \dfrac{0.2}{1-0.8t}$

$E(Y) = 4$

9 $P(X = x) = q^{x-1}p; x = 1, 2, \dots$ where $q = 1-p$

$$G_X(t) = E\left[t^X\right] = \sum_{x=1}^{\infty} q^{x-1}pt^x = tp\sum_{x=1}^{\infty}(qt)^{x-1}$$
$$= tp\{1 + qt + (qt)^2 + \dots\}$$
$$G_X(t) = tp\left(\frac{1}{1-qt}\right)$$
$$G'_X(t) = \frac{p(1-qt)-(-q)(tp)}{(1-qt)^2} = \frac{p}{(1-qt)^2}$$
$$G''_X(t) = \frac{-2(1-qt)(-q)p}{(1-qt)^4}$$
$$= \frac{2pq(1-qt)}{(1-qt)^4}$$
$$= \frac{2pq}{(1-qt)^3}$$
$$G'_X(1) = \frac{p}{p^2} = \frac{1}{p}$$
$$G''_X(1) = \frac{2pq}{p^3} = \frac{2q}{p^2}$$
$$\mu = G'_X(1) = \frac{1}{p}$$
$$\sigma^2 = G''_X(1) + \mu - \mu^2 = \frac{2q}{p^2} + \frac{1}{p} - \left(\frac{1}{p}\right)^2 = \frac{1-p}{p^2}$$

10 $P(X = x) = \dfrac{1}{a}; x = 1, 2, \dots, a$

$$G_X(t) = \frac{t}{a} + \frac{t^2}{a} + \dots + \frac{t^a}{a} = \sum_{i=1}^{a}\frac{t^i}{a}$$
$$G'_X(t) = \sum_{i=1}^{a}\frac{it^{i-1}}{a} \rightarrow \mu = G'_X(1) = \sum_{i=1}^{a}\frac{i}{a} = \frac{1}{a}\frac{a(a+1)}{2} = \frac{a+1}{2}$$

11 a $G_S(t) = \dfrac{1}{a}(1 + t + 3t^2 + bt^3);$

$$G_S(1) = \frac{1}{a}(5+b) = 1 \rightarrow a = 5+b$$
$$G'_S(t) = \frac{1}{a}(1 + 6t + 3bt^2) \rightarrow E(S) = G'_S(1) = \frac{1}{a}(7+3b)$$
$$\frac{7+3b}{5+b} = \frac{5}{3} \rightarrow b = 1\,;\, a = 6$$

b $G''_S(t) = \dfrac{1}{a}(6 + 6bt) = 1 + t \rightarrow G''_S(1) = 2$

$$\text{Var}(S) = G''_S(1) + \mu - \mu^2 = 2 + \frac{5}{3} - \frac{25}{9} = \frac{8}{9}$$

Exercise 1.6B

1 a $G_{X+Y}(t) = e^{-7(1-t)}$

b $X + Y \sim Po(7)$

c If X and Y are two Poisson distributed random variables, means λ and μ, then the random variable $X + Y$ is also Poisson distributed, mean $\lambda + \mu$

2 Let X be the score on first dice; let Y be the score on second dice.

$$G_X(t) = \sum_{i=1}^{6}\frac{1}{6}t^i;\ G_Y(t) = \sum_{i=1}^{8}\frac{1}{8}t^i$$
$$G_{X+Y}(t) = \sum_{i=1}^{6}\frac{1}{6}t^i \times \sum_{i=1}^{8}\frac{1}{8}t^i = \frac{1}{48}\sum_{i=1}^{6}t^i \times \sum_{i=1}^{8}t^i$$
$$= \frac{t^2}{48}\frac{(1-t^6)(1-t^8)}{(1-t)^2}$$

3 a $G_T(t) = \dfrac{t}{6}\left(\dfrac{1}{1-\frac{5}{6}t}\right) \times \dfrac{t}{8}\left(\dfrac{1}{1-\frac{7}{8}t}\right)$

$$= t^2\left\{\left(\frac{1}{6-5t}\right)\left(\frac{1}{8-7t}\right)\right\}$$

b $G_T(1) = 1 \times 1 \times 1 = 1$. $G_T(1)$ is the sum of the probabilities on sample space.

c 14

4 a $G_{X_1+X_2}(t) = 0.36\left(\dfrac{t}{1-0.4t}\right)^2$

b $G_X(t) = 0.36\left(\dfrac{t}{1-0.4t}\right)^2$

c They are equal because both random variables can represent the number of trials up to and including the second success with $p = 0.6$

5 Let U_i be the score on the ith throw.

$$G_{U_i}(t) = \sum_{i=1}^{6}\frac{t^i}{6} = \frac{1}{6}(t + t^2 + \dots + t^6) = \frac{t}{6} \times \frac{1-t^6}{1-t}$$

Let X_T be the total score on the three throws.

$$G_{X_T}(t) = \left(G_{U_i}(t)\right)^3 = \left\{\frac{t}{6} \times \frac{1-t^6}{1-t}\right\}^3 = \frac{1}{216} \times \frac{t^3(1-t^6)^3}{(1-t)^3}$$

6 a $G_{X_T}(t) = (0.6 + 0.4t)^2$

b $B(n = 2, p = 0.4)$

$\mu = 0.8$

$\sigma^2 = 0.48$

7 a $G_{X_T}(t) = (q + pt)^n$ where $q = 1-p$

b $B(n, p)$

$G'_{X_T}(t) = np(q + pt)^{n-1} \rightarrow G'_{X_T}(1) = np;\ \mu = np$

$G''_{X_T}(t) = n(n-1)p^2(q + pt)^{n-2} \rightarrow G''_{X_T}(1) = n(n-1)p^2;$

$\sigma^2 = G''_{X_T}(1) + \mu - \mu^2 = n(n-1)p^2 + np - (np)^2 = np(1-p)$

8 $G_X(t) = E\left[t^X\right]$ and $G_Y(t) = E\left[t^Y\right]$

$$G_{X+Y}(t) = E\left[t^{X+Y}\right] = E\left[t^X \times t^Y\right] = E\left[t^X\right] \times E\left[t^Y\right]$$
$$= G_X(t) \times G_Y(t)$$

Review exercise 1

1 5.16 (3 sf)

2 a 98.5 (3 sf)

b Fairly good agreement between model prediction and actual data suggests hits scattered fairly randomly.

3 a 0.2212 (3 sf)

b $\frac{121}{32}e^{-2.75}(1-e^{-0.25})$

c Mean = 15.4 (3 sf), Variance = 63.7 (3 sf)

d Answer in part **a** shows intervals can't be considered as Bernoulli trials, therefore use of negative binomial is approximate.

4 $P(X>a) = 1 - P(X \le a) = 1 - (1-q^a) = q^a$
(or: $P(X>a) = P(a \text{ failures in } a \text{ trials}) = q^a$)

5 a q^k where $q = 1-p$

b q^k

6 a Mean = 20; variance = 80

b $\bar{X} \sim N(20, 2)$
0.983 (3 sf)

7 a $P(X=x) = \binom{n}{x}p^x q^{n-x};\ x = 0, 1, 2, \ldots, n$ where $q = 1-p$

$$G_X(t) = E\left[t^X\right] = \sum_{x=0}^{n}\binom{n}{x}p^x q^{n-x}t^x = \sum_{x=0}^{n}\binom{n}{x}(pt)^x q^{n-x}$$

$$G_X(t) = q^n + \binom{n}{1}(pt)q^{n-1} + \ldots + (pt)^n = (q+pt)^n$$

b $G_X(t) = (q+pt)^n$

$G_X(1) = (q+p)^n = 1$ because $q + p = 1$

Assessment 1

1 a $k = 10$
$b = 0.4$
$a = 0.15$

b X as $E(X) = 3$ and $E(Y) = 2.65$

2 a $\frac{1}{20}$

b Median is 2
Mode is 2

c $\frac{5}{2}$

d $\frac{21}{20}$ or 1.05

3 a Assume apples are sold individually and at a constant rate.

b i 0.0821
ii 0.5438
iii 0.1088

c 0.1378

4 a $E(U) = 4.5$

$\text{Var}(U) = \frac{35}{12}$ or 2.92 (3 sf)

b $\frac{2}{3}$

5 a 4

b $E(5V+3) = \frac{59}{3}$ or 19.7 (3 sf)

$\text{Var}(5V+3) = \frac{155}{9}$ or 17.2 (3 sf)

6 a $a > 7$ or $a \ge 8$

b $b > 8$ or $b \ge 9$

c 0.6078

d 0.6620

7 a $\frac{3}{4}$

b $\frac{227}{16}$ or 14.2 (3 sf)

8a $\frac{n+1}{2}$

b
$$E(X^2) = \frac{1}{n} + \frac{4}{n} + \frac{9}{n} + \ldots + \frac{n^2}{n}$$
$$= \frac{1}{n}(1+4+9+\ldots+n^2)$$
$$= \frac{1}{n}\left(\frac{n}{6}(n+1)(2n+1)\right)$$
$$= \frac{1}{6}(n+1)(2n+1)$$

$$\text{Var}(X) = \frac{1}{6}(n+1)(2n+1) - \left(\frac{n+1}{2}\right)^2$$
$$= \frac{1}{6}(n+1)(2n+1) - \frac{1}{4}(n+1)^2$$
$$= \frac{1}{12}(n+1)[2(2n+1) - 3(n+1)]$$
$$= \frac{1}{12}(n+1)(n-1)$$
$$= \frac{1}{12}(n^2-1) \text{ as required}$$

9 a 0.0119

b 0.4007

c 0.0368 (3 sf)

10 a i $\frac{27}{64}$

ii $\frac{243}{1024}$

iii $\frac{27}{64}$

b The geometric distribution is memoryless; the probability of '3 more' is the same as of '3 from the beginning'.

11 a $X \sim \text{Negative Binomial}(1, 0.6)$ or $X \sim \text{Geometric}(0.6)$

b $\frac{5}{3}$

c 0.4

12 a $X \sim \text{Negative Binomial}(3, 0.5)$

b 6

c 6

d 0.731 (3 sf)

13 a $G_X(z) = e^{\lambda(z-1)}$

b $G'_X(z) = \lambda e^{\lambda(z-1)}$

$E(x) = G'_X(1) = \lambda$

$G''_X(z) = \lambda^2 e^{\lambda(z-1)}$

$$\text{Var}(X) = G''_X(1) + G'_X(1) - (G'_X(1))^2$$
$$= \lambda^2 + \lambda - \lambda^2$$
$$= \lambda$$

14 a i $P(X=k)=\left(\frac{1}{4}\right)\left(\frac{3}{4}\right)^{k-1}$

ii $G_X(z)=\sum_{x=1}^{\infty} z^x P(X=x)$

$=\sum_{x=1}^{\infty} z^x\left(\frac{1}{4}\right)\left(\frac{3}{4}\right)^{x-1}$

$=\left(\frac{1}{4}\right)z\sum_{x=1}^{\infty} z^{x-1}\left(\frac{3}{4}\right)^{x-1}$

$=\left(\frac{z}{4}\right)\sum_{x=1}^{\infty}\left(\frac{3z}{4}\right)^{x-1}$

$=\left(\frac{z}{4}\right)\left\{\left(\frac{3z}{4}\right)^0+\left(\frac{3z}{4}\right)^1+\left(\frac{3z}{4}\right)^2+\ldots\right\}$

$=\left(\frac{z}{4}\right)\sum_{x=0}^{\infty}\left(\frac{3z}{4}\right)^x$

$=\left(\frac{z}{4}\right)\left(\frac{1}{1-\frac{3z}{4}}\right)$

$=\left(\frac{z}{4}\right)\left(\frac{4}{4-3z}\right)$

$=\frac{z}{4-3z}$

iii $G_Y'(z)=\frac{4}{(4-3z)^2}$

$E(X)=G_Y'(1)=4$

b i $P(Y=k)=\frac{(k-1)}{9}\left(\frac{3}{4}\right)^k$; $k=2, 3, 4, \ldots$

ii $G_Y(z)=\sum_{y=1}^{\infty} z^y P(Y=y)$

$=\sum_{y=1}^{\infty} z^y\times\left(\frac{y-1}{9}\right)\left(\frac{3}{4}\right)^y$

$=\sum_{y=1}^{\infty}\left(\frac{y-1}{9}\right)\left(\frac{3z}{4}\right)^y$

$=\frac{1}{9}\sum_{y=1}^{\infty}(y-1)\left(\frac{3z}{4}\right)^y$

$=\frac{1}{9}\sum_{y=1}^{\infty}(y-1)\left(\frac{3z}{4}\right)\left(\frac{3z}{4}\right)^{y-1}$

$=\frac{3z}{36}\sum_{y=1}^{\infty}(y-1)\left(\frac{3z}{4}\right)^{y-1}$

$=\frac{3z}{36}\sum_{y=0}^{\infty}y\left(\frac{3z}{4}\right)^y$

$=\frac{3z}{36}\left(\frac{\left(\frac{3z}{4}\right)}{\left(1-\frac{3z}{4}\right)^2}\right)$

$=\frac{3z}{36}\left(\frac{\frac{3z}{4}}{\left(1-\frac{6z}{4}+\frac{9z^2}{16}\right)}\right)$

$=\frac{3z}{36}\left(\frac{12z}{16-24z+9z^2}\right)$

$=\frac{3z}{36}\left(\frac{12z}{(4-3z)^2}\right)$

$=\frac{z^2}{(4-3z)^2}$

iii $G_Y'(z)=\frac{2z(4-3z)^2-z^2\times 2(4-3z)\times -3}{(4-3z)^4}$

$=\frac{2z(4-3z)+6z^2}{(4-3z)^3}$

$G_Y'(1)=\frac{2(1)+6}{1}$

$=8$

c For equal probabilities the negative binomial distribution is 'memoryless'; i.e. waiting for one failure and then waiting for two more failures is the same as waiting for three failures from the beginning.

15 a $r=2,\ p=\frac{4}{5}$

b $E(X)=\frac{5}{2}$

$\text{Var}(X)=\frac{5}{8}$

c $\frac{608}{625}$ (= 0.973 (3 sf))

16 a $p=0.25$

b Mean = 4, Variance = 12

c Mean = 2.5, Variance = 3.75

The mean and variance of Y are both smaller than the mean and variance of X.

d A strictly decreasing function is one whose derivative is always negative.

$\frac{d}{dp}\left(\frac{1}{p}\right)=-\frac{1}{p^2}<0$

$\frac{d}{dp}\left(\frac{1-p}{p^2}\right)=\frac{d}{dp}\left(\frac{1}{p^2}-\frac{1}{p}\right)=\frac{p-2}{p^3}<0$ as $0<p<1$

As p increases, both mean and variance decrease, so the claim is wrong.

17 a $X+Y\sim N(10, 32)$

$X-Y\sim N(0, 32)$

b i 0.309 (3 sf)

ii 0.227 (3 sf)

iii 0.188 (3 sf)

Exercise 2.1A

1 a $H_0: \lambda = 5.31$, $H_1: \lambda < 5.31$
  b 53
2 a 21 and 41
  b $3 \times 9 = 27$ is between the critical values so the null hypothesis is accepted.
3 a $H_0: \lambda = 8.3$, $H_1: \lambda \neq 8.3$
  b 0.0885
  c Since the p-value is larger than half the significance level (2-tailed test) the null hypothesis is accepted.
4 a 163
  b 0.04505
  c Since the p-value is less than the significance level the null hypothesis is rejected.
5 a 72
  b 0.0482
6 a 0.0568
  b Since the p-value is larger than the significance level the null hypothesis is accepted.
  c 0.0469

Exercise 2.1B

1 a 34 and 63
  b The result is outside of the critical region so the null hypothesis is accepted. You conclude that the average rate of phone calls to the call centre is reasonably estimated at 6.8 per 10 minutes.
  c A type I error is when a true null hypothesis is falsely rejected. It occurs when a result in the critical region is obtained.
  d 0.0429 (3 sf)
2 a 2122
  b As $2098 < 2122$ the null hypothesis is accepted. You conclude that the average number of points scored per game has not increased.
3 a 0.0838 (3 sf)
  b Since the p-value is larger than the significance level, the result is not significant and the null hypothesis is accepted. There is insufficient evidence to suggest that there are fewer accidents per month at the sports centre.
4 a 0.0420
  b The p-value is smaller than the significance level. There is sufficient evidence to reject the null hypothesis. You conclude that the average rate of customers coming into the book shop has increased.
5 a 0.1174
  b The p-value is larger than half the significance level (2-tailed test). There is insufficient evidence to reject the null hypothesis. There is insufficient evidence that the number of light bulbs which burn out each month has changed.
6 a 3 and 13
  b 0.0213 (3 sf)
7 a i P(type I) = 0.0777 (3 sf)
    ii P(type I) = 0.0162 (3 sf)
    iii P(type I) = 0.00185 (3 sf)
  b i P(type II) = 0.772 (3 sf)
    ii P(type II) = 0.916 (3 sf)
    iii P(type II) = 0.980 (3 sf)

Exercise 2.2A

1 a 6
  b 0.335 (3 sf)
  c 30
2 a $X \geq 30$
  b $X \geq 60$
  c $X \geq 300$
3 a 0.0798 (3 sf)
  b 0.00573 (3 sf)
  c 0.000411 (3 sf)
4 a $X \geq 6$
  b 2.5, 3.75
  c This is in the critical region so there is sufficient evidence to reject the null hypothesis. You conclude that the probability of a success is less than 0.4
5 a 12
  b 0.0674 (3 sf)
  c This is greater than the significance level so there is insufficient evidence to reject the null hypothesis. You conclude that the probability of rolling a 12 is $\frac{1}{12}$
6 $X \leq 5$

Exercise 2.2B

1 a The probability of success is unlikely to be the same with each person as there is no range of numbers specified.
  b $H_0: p = 0.2$, $H_1: p < 0.2$
  c $0.0115 < 0.05$ There is sufficient evidence at the 5% level to suggest that the student guesses correctly with probability less than 0.2
2 a The probability of success is equal each time and the outcomes are independent.
  b $H_0: p = \frac{1}{13}$, $H_1: p \neq \frac{1}{13}$
  The p-value of the statistic is 4.41%, which is less than the significance level. There is sufficient evidence to suggest that this is not a regular pack of cards.
3 a The probability of $X = 1$ is larger than the significance level so no result will be sufficient evidence.
  b 0.025
4 a There is a constant probability that success and events are independent, even if they have to stop digging before a success occurs.
  b 0.00320 (3 sf)
5 a If q is not smaller than the significance level, then even $P(X = 1)$ is larger than the significance level.
  b As q decreases, the variance increases. The difference between the two critical values, therefore, increases.

Exercise 2.3A

1 a $X \geq 7$
  b 0.0989
  c This is inside the critical region, so the result is significant. There is sufficient evidence to reject the null hypothesis; you conclude that the true value of p is less than 0.32
2 A type I error is when a null hypothesis is rejected when it is in fact true. A type II error is when a null hypothesis is accepted when it is in fact false.
3 a 0.05
  b 0.518
  c 0.999
4 a $X \geq 12$
  b 0.0448
  c 0.730
  d 0.270
5 a P(type II) = 0.771, power = 0.229
  b P(type II) = 0.416, power = 0.584
  c P(type II) = 0.021, power = 0.979

6 a Power is 0.852

b The power is symmetric about the hypothesis mean for a Normal distribution.

c As the true mean gets further from the hypothesis mean, the smaller the probability of obtaining an outcome not in the critical region, so the probability of making a type II error decreases.

Exercise 2.3B

1 a Critical value = 1.07, P(type II error) = 0.350, power = 0.650

b Critical value = 1.28, P(type II) = 0.491, power = 0.509

c Critical value = 1.70, P(type II) = 0.745, power = 0.255

2 a i $X \geq 4$

ii 0.799

iii 0.201

b i $X \geq 5$

ii 0.916

iii 0.084

3 a 8

b P(type II error) = 0.978, power = 0.022

c 800

d You should expect the power to be much closer to 1 than for the 20-trial test, due to the larger sample size.

4 a The treasure hunter does not want to miss any valuables.

b If the hunter digs a little and finds nothing valuable then they can keep trying other locations.

5 a $X = 0$

b 0.0143

c P(type II error) = 0.9999, power = 0.0001

6 a $X \geq 5$, 0.249

b $X \geq 8$, 0.399

c $X \geq 11$, 0.492

7 a Since 5 games are played, 1.2 goals per game leads to 6 goals per 5 games. The alternative is > 6 because the team thinks they might be scoring more goals.

b Critical region is $X \geq 11$, 0.0426

c 0.774

d They probably can't tell the difference between an extra goal every 5 away matches.

e An extra goal will change the odds enough that the company could lose a lot of money making bad predictions.

Exercise 2.4A

1 a 8

b 9, 16.9

c The X^2 statistic is 16.25. This is less than the critical value so the null hypothesis is accepted.

2 a 2.91%

b 13.22%

c 2.20%

d 8.39%

3 i a 4, 7.78

b The X^2 statistic is 23.6. The statistic is larger than the critical value so there is sufficient evidence to reject the null hypothesis.

ii a 5, 9.24

b The X^2 statistic is 4.10. The statistic is smaller than the critical value so there is insufficient evidence to reject the null hypothesis.

iii a 6, 10.6

b The X^2 statistic is 8.78. The statistic is smaller than the critical value so there is insufficient evidence to reject the null hypothesis.

4 a 8.78

b 8.42

c Both people conclude that their model is a good fit for the situation.

d Both models are a good fit for the situation and neither one is likely to be a perfect description. There is no paradox.

Exercise 2.4B

1 a 0.6

b

Outcome	0	1	2	3	4
Expected frequency	1.92	11.52	25.92	25.92	9.72

c

Outcome	0–1	2	3	4
Expected frequency	13.44	25.92	25.92	9.72

d 6.59

e 2, 5.991

f The statistic is larger than the critical value so you reject the null hypothesis. You conclude that the binomial distribution is not a good model for the situation.

2 a $\lambda = 2$

Outcome	0	1	2	3	4	5+
Expected frequency	6.77	13.53	13.53	9.02	4.51	2.64

b Grouping the 4+ outcomes together, the X^2 statistic is 2.93. For $5 - 1 - 1 = 3$ degrees of freedom at the 5% level the critical value is 7.815. The statistic is smaller than the critical value so the null hypothesis is accepted. You conclude that the Poisson distribution is a suitable model for the situation.

3 The expected frequencies must be 13 to sum to the same as the observed frequencies. There are $5 - 1 = 4$ degrees of freedom so the critical value at the 10% level is 7.779. The X^2 statistic is 11.85, which is larger than the critical value. The null hypothesis is rejected and you conclude that the discrete uniform distribution is not a good description of the situation.

4 a 1.6

b 0.625

r	1	2	3	4	5	6+
Expected frequency	93.75	35.16	13.18	4.94	1.85	1.11

c Grouping frequencies together,

r	1	2	3	4+
Expected frequency	93.75	35.16	13.18	7.91
Observed frequency	84	47	15	4

There are $4 - 1 - 1 = 2$ degrees of freedom. The critical value at the 5% significance level is 5.991. The X^2 statistic is 7.18. The statistic is larger than the critical value so there is sufficient evidence to reject the null hypothesis. You conclude that this model is not a reasonable way to describe the number of attempts required to pass a driving test.

5 a

x	1	2	3	4	5+
Expected frequency	21	14.7	10.29	7.203	16.807

b $5 - 1 = 4$ degrees of freedom. The critical value at the 10% significance level is 7.779. The X^2 statistic is 9.7536.

The statistic is bigger than the critical value so there is evidence to reject the null hypothesis. You conclude that the model is not suitable for the variable.

Exercise 2.5A

1 a

Expected	A	B	C
W	54.6	88.0	11.4
X	14.2	22.9	3.0
Y	70.2	113.1	14.6

b

X^2	A	B	C
W	0.10	0.18	3.59
X	3.64	3.66	1.30
Y	0.33	1.53	4.76

X^2 Total	19.0823

c 4, 9.49

d 0.000757

e The test statistic is larger than the critical value (or the p-value is less than the significance level), so there is sufficient evidence to reject the null hypothesis.

Using Yates:

b

X^2	A	B	C
W	0.07	0.14	3.05
X	3.15	3.27	0.72
Y	0.26	1.41	4.21

X^2 Total	16.274

d The p-value is 0.00267

2 a

Expected	A	B	C	D
W	12.4	12.1	14.9	15.6
X	10.3	10.1	12.4	13.1
Y	13.0	12.8	15.7	16.5
Z	13.3	13.0	16.0	16.8

b

X^2	A	B	C	D
W	1.07	0.07	0.08	2.04
X	0.53	0.81	1.59	1.17
Y	1.20	2.61	0.108	0.02
Z	2.09	0.31	0.26	0.09

X^2 Total	14.04275

c 9, 16.9

d 0.120839

e The test statistic is smaller than the critical value (or the p-value is larger than the significance level), so there is insufficient evidence to reject the null hypothesis.

3 a

Expected	A	B	C
W	74.4	68.5	78.1
X	65.6	60.5	68.9

b

X^2	A	B	C
W	0.29	1.94	0.61
X	0.33	2.20	0.69

X^2 Total	6.055767

c 2 , 5.99.

d 0.04842

e The test statistic is larger than the critical value (or the p-value is less than the significance level), so there is sufficient evidence to reject the null hypothesis.

4 a

Expected	A	B
W	80.6	113.4
X	39.4	55.6

b

X^2	A	B
W	1.25	0.89
X	2.56	1.82

X^2 Total	6.52749

c 1 , 3.84.

d 0.010623

e The test statistic is larger than the critical value (or the p-value is less than the significance level), so there is sufficient evidence to reject the null hypothesis.

5 a

E_i	0–1	2–3	4+
Brown	8.8	12.7	13.5
White	14.2	20.3	21.5

b

X^2	0–1	2–3	4+
Brown	0.916	0.007	0.479
White	0.57	0.005	0.30

c 2.27802.

d 2, 0.320139

e The p-value is larger than the significance level so there is insufficient evidence to reject the null hypothesis.

6 a

E_i	Chicken	BLT	Ham & Cheese	Egg Mayo
Orange	26.9	28.4	20.4	8.4
Apple	23.4	24.6	17.7	7.3
Water	23.7	25.0	17.9	7.4

b

X^2	Chicken	BLT	Ham & Cheese	Egg Mayo
Orange	0.622	0.005	0.091	0.668
Apple	0.822	0.768	0.027	0.074
Water	0.004	0.636	0.237	0.362

X^2 Total	4.31525

c The statistic is less than the critical value so there is insufficient evidence to reject the null hypothesis.

Exercise 2.5B

1 a The total X^2 statistic is 18.91. For 4 degrees of freedom at the 5% significance level the critical value is 9.49. The test statistic is larger than the critical value so the result is significant at the 5% level. There is sufficient evidence to reject the null hypothesis.

b The largest contribution is from people with blue eyes who say blue is their favourite colour significantly more than expected. The number of people with green eyes whose favourite colour is red and those with brown eyes whose favourite colour is yellow is very close to what's expected.

2 a Using Yates' correction the total X^2 statistic is 2.86. At the 5% significance level with one degree of freedom the critical value is 3.84. The statistic is lower than the critical value so there is insufficient evidence to reject the null hypothesis.

b Although in the observed data there are more brown stones on the beach than by the river, the numbers are not dissimilar enough from the expected values to suggest there is any association.

3 a The total X^2 statistic is 27.9. For 6 degrees of freedom at the 5% significance level the critical value is 12.6. The test statistic is larger than the critical value so the result is significant at the 5% level. There is sufficient evidence to reject the null hypothesis.

b The association suggests that people who play the different tribes choose their weapons accordingly and not independently. The largest contributions are from people who play as Huns using far fewer Axes than expected and far more Bows than expected. People who play as Vikings use more Axes than expected and fewer Bows than expected. Spears are the most popular weapon but they occur around the expected frequency for each tribe.

4 a With Yates' correction the test statistic is 3.839. The critical value with 1 degree of freedom at the 10% level is 2.706. The test statistic is larger than the critical value so there is sufficient evidence to reject the null hypothesis. You conclude that there is association between choice of starter and main meal.

b No individual contribution is very large or small but together they suggest association. Pasta + Soup and Pizza + Salad go together slightly more than you would expect, and the other pairs go together slightly less than you would expect.

5 a The total X^2 test statistic is 2.59. The critical value with 2 degrees of freedom at the 1% level is 9.21. The test statistic is less than the critical value so there is insufficient evidence to reject the null hypothesis. You conclude that there is no association between child gender and monster colour preference.

b The numbers of boys and girls who prefer the yellow monster match the expected numbers well. There are no large contributions to the test statistic.

6 Check correct for Ex 2.5A Q4 and Ex 2.5B Qs 2 and 4

Review exercise 2

1 a $H_0: \lambda = 1.9, H_1: \lambda > 1.9$

b 0.0332 (3 sf)

c The p-value is less than the significance level so there is sufficient evidence to reject the null hypothesis.

2 a 13

b 3.11% (3sf)

3 a 5

b 7 is outside the critical region so there is insufficient evidence to reject the null hypothesis. You conclude that there are not fewer accidents per month than previously, and that 1.3 is a reasonable estimate of the rate of accidents per month.

4 a $X \geq 71$

b 81.7%

5 a 8

b 2.08%

c This is less than half the significance level. There is sufficient evidence at the 5% level to suggest that the probability of rolling an 8 is not $\frac{1}{8}$

6 a 1.1

x	0	1	2	3	4+
E	26.63	29.29	16.11	5.91	2.06

b The X^2 contributions sum to 15.10. This has a p-value of 0.000527, so the result is significant. There is sufficient evidence to suggest that the Poisson distribution model is not a good fit.

7 a 0.63

b

x	0	1	2	3	4	5	6
E	0.64	6.55	27.89	63.33	80.87	55.08	15.63

c

x	0–1	2	3	4	5	6
E	7.19	27.89	63.33	80.87	55.08	15.63

d 4, 9.488

e 4.530

f There is insufficient evidence to reject the null hypothesis. You conclude that it is reasonable to model the population as a binomial distribution with some parameter, p.

8 a

E_i	A	B	C
X	100.0	68.7	40.3
Y	105.7	72.6	42.6
Z	104.3	71.7	42.1

b

E_i	A	B	C
X	0.091	0.077	0.701
Y	0.895	1.775	0.06
Z	0.432	2.603	1.148

7.78 (3 sf)

9 12.96

10 a 8

b 15.5 (3 sf)

11 The p-value is 6.97% (3 sf). This is larger than the significance level so there is insufficient evidence to reject the null hypothesis. You conclude that there is no association between height and birthplace of UK adults.

12 a

E_i	A	B	C	D
Red	86.4	85.9	67.2	84.5
Blue	67.7	67.3	52.7	66.2
Silver	30.9	30.8	24.1	30.3

b

X^2	*A*	*B*	*C*	*D*
Red	2.151	0.042	0.777	0.240
Blue	2.385	0.166	0.350	2.086
Silver	0.028	0.894	0.357	1.739

11.22

c The test statistic is smaller than the critical value so there is insufficient evidence to reject the null hypothesis. You conclude that there is no association between the public's preference for the four car models and their choice of colour. For model *A* there is positive association with Red and negative association with Blue. For model *D* there is positive association with Blue and negative association with Silver. Models *B* and *C* appear to have colour preferences in roughly the expected proportion.

Assessment 2

1 a H_0: $\lambda = 6.4$, H_1: $\lambda < 6.4$

b Under H_0, $P(X \le 2) = 0.046$ (3 dp)
For a 5% one-tailed test, reject the null hypothesis; evidence suggests that the number of flaws has been reduced.

c Rejecting the null hypothesis when it is true.
$1 - 0.046 = 0.954$

2 a H_0: $\lambda = 12 \times 6 = 72$, H_1: $\lambda > 72$
Under H_0, $P(X \ge 90) = 0.022$ (3 dp)
For a 5% one-tailed test, reject the null hypothesis; evidence suggests that the number of calls has increased.

b 0.0470
This is the probability that the null hypothesis is rejected given that it is true.

3 a $X \sim$ Geometric (p). Assume independence of outcomes for each lens.
H_0: $p = 0.23$, H_1: $p \ne 0.23$

b 0.160 (3 dp)

c No reason to reject H_0; evidence suggests no change to proportion rejected.

4 a H_0: $\lambda = 2.7$, H_1: $\lambda > 2.7$

b $X \ge 7$

c 0.762 (3 dp)
Power = 0.238

5 a

x	0	1	2	3	4	>4
P(*X* = *x*)	0.1827	0.3106	0.2640	0.1496	0.0636	0.0295
Expected frequency	6.577	11.180	9.503	5.385	2.289	1.066

Poisson distribution, mean 1.7, because of random events.

b Pool last three cells together:

x	0	1	2	≥ 3
f(*x*)	8	12	10	6
Expected frequency	6.577	11.180	9.503	8.740

$\chi^2 = 1.25$ (2 dp)
$\nu = 3$
The critical region for this test is $\chi^2_3 > 7.815$
No reason to reject null hypothesis; evidence suggests that calming measures had no impact on accident rate.

6 a

x	1	2	3	4	5	6	> 6
P(*X* = *x*)	0.4	0.24	0.144	0.0864	0.05184	0.0311	0.04666

b Pool last three cells together:

x	1	2	3	4	≥5
Observed frequency	13	7	8	6	11
Expected frequency	18	10.8	6.48	3.888	5.832

$\chi^2 = 8.81$ (2 dp)
$\nu = 4$
The critical region for this test is $\chi^2_4 > 7.779$
Reject null hypothesis; evidence suggests that $p \ne 0.4$

7 a

		Result			
		Fail	Pass	Distinction	Total
Class	A	7.7	17.6	7.7	33
	B	6.3	14.4	6.3	27
	Total	14	32	14	60

b $\chi^2 = 0.19$ (2 dp)
$\nu = 2$
The critical region for this test is $\chi^2_2 > 5.991$
No reason to reject teacher's belief.

8 Expected frequencies are:
Type A: 45.868, 17.132; type B: 23.298, 8.702; type C: 13.833, 5.167
$\chi^2 = 0.146$
$\nu = 2$
The critical region for this test is $\chi^2_2 > 5.991$
There is no reason to reject the hypothesis that weekdays and weekends result in the same proportion of the different errors.

Index